Farmers
for the
Future

The author with the Lowensteins—potential farmers for the future.

Farmers
for
the
Future

DAN LOOKER

Iowa State University Press / Ames

Dan Looker is business editor of *Successful Farming*. A journalist since 1976, he was named Agricultural Journalist of the Year in 1990. In 1994-95 Looker served as president of the National Association of Agricultural Journalists.

♾ Printed on acid-free paper in the United States of America

First edition, 1996

Portions of this book and most of the photographs first appeared in *Successful Farming*. They are copyrighted © by the Meredith Corporation and appear here by kind permission of the Meredith Magazine Group. Other portions have appeared in *Farmers for the Next Century,* the proceedings of The First National Conference for Beginning Farmers and Ranchers.

Library of Congress Cataloging-in-Publication Data

Looker, Dan
    Farmers for the future/Dan Looker.
       p.   cm.
    ISBN 0-8138-2383-8
      1. Farmers—United States. 2. Agriculture—Economic aspects—United States. I. Title.
    HD8039.F32U653   1996
    338.1'0973—dc20   95-37275

To
Allen Prosch,
one of the
pioneers
of the renewal
of farming
in America

Contents

Farmers Helping Themselves

Acknowledgments

I t's wonderful to work with those who share your enthusiasm for a project. Without the vision of *Successful Farming* editor Loren Kruse and the support of managing editor Gene Johnston, I doubt that this book would have been written. Both men share my interest in helping young people find a realistic, practical way to get started in farming.

Other writers at *Successful Farming* also offered ideas and suggestions. Rich Fee and Rod Fee helped find farmers to speak at our first two national conferences for beginning farmers and ranchers, Farmers for the Next Century. Part of this book has its roots in those conferences. They would not have taken place, either, without financial support from John Deere Life Insurance Company, John Deere Credit, and the Farm Credit Services of the Midlands and of Missouri and Illinois.

Thanks are due to Rick Jost, business editor of *The Des Moines Register*, where, as a reporter, I first explored the loss of America's youngest farmers in a series of articles published in that newspaper in 1990. Rick gave me valuable time to write that series. I'm also indebted to Geneva Overholser, former editor of *The Register*, whose support of coverage of agriculture has been steadfast when many newspapers have forgotten this important part of the nation's economy and heritage.

I am especially grateful to the staff at the Center for Rural Affairs—a remarkable rural brain trust tucked away in Walthill, Nebraska—in particular to Marty Strange, its brilliant co-founder, and to Nancy Thompson, its dedicated attorney. I began my career in agricultural journalism by working for the Center in 1976 and 1977. Since then, the Center has become a leader in beginning farmer issues.

I also thank the many farmers who shared part of their lives for this book. And I am indebted to the authors and scholars who have allowed me to briefly cite their work on public policy issues that affect beginning farmers. And to Bill Silag, former editor-in-chief of Iowa State University Press, who got this book started.

I owe my affection for rural America to my mother, Ellen Castleman Looker, who spent her early childhood on an Oregon berry farm and a subsistence ranch in central Washington. And I owe it to my wife's parents, Miles and Genevieve McCullough, who ran a cattle ranch near North Platte, Nebraska, for more than a half century. They exemplify the best qualities of farm and ranch people.

Thanks, too, to my wife Joan and my children for their patience during my struggles with writer's block and deadlines.

Introduction

This book is about hope and renewal. It's about the kind of personal energy and character that built a great nation, the kind of drive that pushes today's young farmers. This book is about and for those young people of America who are beginning farming when it may be more difficult than at any time in history. It's also for anyone who wonders why it's still vital that independent, family-run farms survive into a new generation.

For many reasons, offering hope to young people who want to start out in production agriculture seems to scoff at conventional wisdom in the late 20th century. At first, farming without a lot of financial help from parents or relatives does, indeed, seem hopeless.

Every five years, the U.S. Commerce Department publishes a detailed *Census of Agriculture*. The last two, taken in 1987 and 1992, showed a steep drop in numbers of young people in farming and ranching.

Behind the numbers are tough realities. Agriculture in the Midwest and Plains—the heart of traditional family farming—is going through wrenching change. The hog industry, which offered an inexpensive early foothold in agriculture to generations, may be falling under control of big, vertically-integrated companies. Profit margins from growing corn, soybeans and wheat seem tighter every year as the federal government cuts price supports, as export markets grow more complex and uncertain, and as technology turns once poor countries such as India into modest grain exporters. Meanwhile, the cost of buying enough land and machinery to support a family by growing only corn and soybeans is now about $1 million. Even most Americans at retirement don't have that.

To top it off, older farmers and ranchers feel pressure from environmentalists, animal rights activists, rising property taxes, tougher financial reporting requirements at banks—a swarm of management headaches. It makes them wonder why anyone wants to invest so much for the privilege of working outdoors in mud, blizzards and drought to earn a modest living with no paid vacation or health benefits.

Steve Hopkins, who began a tiny, rented dairy farm in northeast Iowa with his wife, Sara Andreasen, puts it well: "I could have gone to law school or medical school or veterinary school ... but that's not what I wanted to do. I was more interested in farming. Still, I think that farming would have been the most difficult path to take of any of those occupations."

But Steve and Sara are farming, even though Sara has another job to help support a young family. Their business is profitable. They're buying the cows that they first rented. And they are buying a small amount of land.

In spite of the current pessimism in agriculture, there may be thousands of Steves and Saras farming at the turn of the century—young people who are well educated, who worked at other jobs first to save up a grubstake, and who moved onto the land only after careful planning and training.

The most hopeful sign of a budding rural renaissance came in March of 1994, when The First National Conference for Beginning Farmers and Ranchers, dubbed Farmers for the Next Century, played to a sold-out crowd of nearly 400 in Omaha, Nebraska. It was organized by *Successful Farming* magazine of Des Moines, Iowa, the nation's largest subscription-based farm magazine, and by the Center for Rural Affairs, a nonprofit farm advocacy group in Walthill, Nebraska, that has worked for years to prod the government to help young farmers. To keep registration fees low, the organizers sought financial support from agricultural businesses. It was a tough sell, given the general skepticism about entering farming. But two entities stepped forward, John Deere Life Insurance Company, which offered estate planning workshops through Deere and Company machinery dealerships, and Farm Credit Services of Omaha, part of the nationwide Farm Credit System, a farmer-owned lending cooperative.

After a slow start in getting backing, the organizers had about two months to publicize the conference. Yet, it sold out in less than six weeks. It could have been twice as big if a larger conference site had been booked. Even though the conference was crowded and speakers at times were hard to hear, the men and women who attended were enthusiastic. As expected, most came from Nebraska and Iowa, but the meeting drew from Ohio, Texas and California. The following year, with additional help from another business backer, we organized a bigger and better conference in Columbia, Missouri. It, too, was sold out.

The conferences also brought older farm owners who wanted to find young people to work into their businesses. In rural areas where many farmers have gray hair, interest in farming by young people isn't obvious. But this national response to the conference suggests no shortage of young people eager for the hard work of farming and ranching.

Here's another hopeful sign: Older farmers nearing retirement, without children who want to farm, are starting to worry about whether their own work will be carried on—or just be swallowed by the biggest farm in the county. Thoughtful farmers worry, too, about an impending exodus from farming if too many people retire at once and few take over. That will devastate small-town shops, schools, and churches already suffering from rural attrition. Some of those concerned farmers are directors of Farm Credit Services of Omaha, which operates in the Iowa, Nebraska, South Dakota and Wyoming. In 1994, those farmers voted to set aside $20 million of the Omaha bank's loan funds for younger farmers' operating loans—to help plant crops, buy machinery or raise livestock. It's not a giveaway. Each loan proposal must show a realistic way to be repaid. But the bank will require less net worth than it normally does of established farmers.

Farmers like Dwight and Sally Puttmann of Kingsley, Iowa, have taken matters into their own hands. They went to Iowa State University in 1987 to recruit a student to work on their farm and gradually take over its management.

Other older farmers and established farm organizations may follow these examples to offer opportunities to young people. Cooper Evans, a respected former Republican congressman from Iowa and White House agricultural advisor to President George Bush, thinks the need for more young farmers will grow so obvious that, "a rather broad coalition of strange bedfellows" will pressure the federal government to help young farmers.

A final reason for hope for young farmers has to do with the fact that farming goes through business cycles just like the rest of the economy. Sometimes, before cycles shift, the experts are dead wrong about the future. That may be true of today's dire outlook for young farmers.

Not so long ago, in the 1970s, agriculture was caught up in a euphoria driven by big grain sales to the former Soviet Union and by inflation that made farmland seem as secure and safe as buying gold. "They're not making any more of it," everyone said, and lenders made it easy to buy, with as little as 10 percent down. In the late 1970s and early 1980s, respected agricultural economists predicted an ever higher spiral in land prices. The opposite happened. Agriculture went through a depression that saw land prices fall by 65 to 72 percent in the 1980s. Young farmers or parents who borrowed to buy land just before the market crashed were washed out of business, even though most were skilled and hardworking. It took an act of Congress to keep the Farm Credit System solvent. Lenders in general became much more cautious, often requiring a 50 percent down payment on a farm mortgage. Today, the pessimism left from the decade of the 1980s may be too great. Maybe, just maybe, the enthusiastic crowd of young people in Omaha in 1994 was a bellwether of rural renewal. Or, maybe it was just an aberration in the long decline of farm numbers now more than a half century old. It's too soon to tell.

What was clear at the end of the 20th century is that the world's economy was changing in unexpected ways. In 1994, *The Economist*, the 151-year-old conservative British news weekly, published a cover editorial on "The Fall of Big Business." As trade barriers fell and technology made the cost of manufacturing less expensive, smaller companies were outcompeting the big multinational companies that experts had expected to profit from increased trade. A company that seems small to *The Economist* is likely to be much bigger than most farm businesses, and agriculture very well could continue concentrating into fewer and fewer hands. But the troubles of IBM and other giants were proof again that the experts can be wrong. The economic powerhouses of shared monopoly were no more eternal than the political grasp of the Kremlin in the now-defunct Soviet Union and of racial apartheid in South Africa.

This book is divided into three sections that can stand alone. The first deals with the economic and political reality facing young farmers today. It's for the general public and those less familiar with farming as well as young people from farms. It may be the most discouraging section, but it's important for anyone going into a business to know its risks. The second section describes the few government and private programs that exist to help young farmers. With the deficit-cutting mood in Washington, the federal programs described here may be out of date soon after this book is published. Fortunately, some state and private programs are springing up that could help make up for Washington's shortsightedness. The last section describes what farmers have done on their own to get started, or to help younger people work into their operations later in life. This is the most exciting and encouraging part of the book.

This book is written with the goal of helping to create more opportunity for more people to enter agriculture. Only about a third of Iowa's existing farmers expect their children to take over their farms, according to a recent poll by Iowa State University sociologist Paul Lasley. That suggests a big need to bring new kids onto the section—the square mile of land that's the block in Midwest rural neighborhoods.

For those bold young people who want to farm, this book is a first step toward a lifetime of learning about agriculture—and harvesting its true richness of self-reliance, a sense of place and community, and a love of the natural resources we have been given.

Roots
of Decline,
Reasons
for
Renewal

CHAPTER 1

Pat and Julie Steffen and their sons, Tom and Robert, in 1991, the year they bought their northeast Nebraska farm. Photo by Russ Munn, courtesy of *Successful Farming*.

Boom and Bust: A Sobering Background for Today's Young Farmers

The two-story farmhouse cupped between waves of grassy hills in northeast Nebraska looks peaceful and secure in the green of late May. Here Pat and Julie Steffen farm and are raising their three young children. The house near the town of Fordyce is on a century-old farmstead. Pat, the youngest of ten children, marks the fourth generation of his family to work and live in this German Catholic community of modest farms.

This is the western edge of the Corn Belt of the United States, one of the richest natural resources on earth. Although corn is grown in many parts of the nation, the best soil and climate for the crop stretches from Ohio into Nebraska. In central Nebraska, near the Oregon Trail that brought European settlers to the state, irrigated fields often yield a bounty of 200 bushels of corn an acre.

The Steffen farm isn't in one of the high-rent Corn Belt neighborhoods, though. It's a few miles from where the Missouri River bends northward into South Dakota. West of here, stands of hardy burr oak along river bluffs shift to pine and cedar. The uplands change from scattered, irrigated cornfields to vast grasslands grazed by cattle. The Steffen farm, located in this zone between Corn Belt and cattle country, gets only about 21 inches of rainfall a year. The best land in the area "will produce about 90 bushels an acre when everything else dries out," Pat says. It sells for $1,000 an acre, tops. That's well under the $2,000 an acre or more that prime Corn Belt land was bringing in the mid-1990s.

The quiet scene at the Steffen house gives little hint of the struggle that went into keeping this farm. Like many capable, hardworking American farm families, the extended Steffen clan was caught in a financial and economic quake that rocked rural America in the 1980s. They lost their farm. Pat, now 33, and Julie, 35, slowly regained it. The Steffens' experience, multiplied by thousands across the nation, left a legacy of cautious bankers that even today makes it harder for young people to borrow to start farming. But Pat and Julie's story shows the resilience of a new generation. It is evidence that in spite of this cautious new financial climate, young families can get started on their own with modest resources. And, unlike farmers who joined the "back-to-the-land" movement during an inflationary boom in the 1970s, Pat and Julie's farm may have a better chance of surviving. They have borrowed little and expect no easy route to prosperity.

Pat's father, Andy, went into debt in the late 1970s to help one of Pat's older brothers get started in farming on his own.

"One of the boys farmed 14 or 15 miles west of here, so we invested in machinery to help him out," Andy recalls, looking wistfully out the kitchen window of the house where he was born and where Pat and Julie now live. Pat's older brother had started farming in 1975. Late in the decade, his father bought a used combine for $25,000, upgrading to a more expensive model three years later. Expanding beyond the 400 acres that Andy Steffen and his wife, Rita, owned, the father-son business rented more cropland. The farm grew to 1,000 acres. The Steffens leased a $65,000 insulated building with continuous flow waste gutters that they used to feed their hogs to market weight.

By the early 1980s, the Federal Reserve Board's efforts to stop inflation by raising interest rates hit many farmers like the Steffens with a vengeance. Interest rates rose to 18 percent or higher.

"Dad expanded right when inflation got real high," Pat recalls. "Everything unraveled at the seams."

Other misfortunes compounded the high interest rates. The hog building proved overrated by its salesmen. The electric bill for its ventilation fans ran $400 a month and the hogs didn't gain weight fast enough to offset the high overhead costs. In the winter of 1980-81—Pat's first year out of high school—hog prices fell to a disastrous 28 cents a pound. Pat and another brother were working on the home place by then, while the family worked with his older brother who was farming to the west.

Their banker offered to help turn things around by having the banker's brother-in-law buy some bred cows to pasture. They were malnourished and 16 of 40 died. "We could have bought a rendering plant with the cows we lost," Andy says, able to laugh now about a painful time for the family.

"There was some days when a guy didn't know which way to turn, I'll tell you that," he confides.

The bank carried the farm with more short-term loans "until, finally, we reached the point of no return and we had a farm sale," Andy says.

Although Andy says he made plenty of mistakes on his own, there is still some bitterness in Rita's voice when she recalls the sale in December 1984. "Losing the farm was almost like a death in the family. When it happens, you're losing your life's work."

"We made one mistake that year," she says. "We harvested the corn and we had that expense. We should have let the banker do it."

The sale of most of their machinery and livestock repaid the banker. Another bank which held the mortgage on the leased hog building acquired the deed to the land at a sheriff's sale in 1985. The Steffens didn't get enough from the machinery sale to pay all of their suppliers, though, so Andy cashed in his $150,000 life insurance policy. "We got them all paid," Rita says proudly.

For Pat, who married Julie in August 1984, the loss hit hard. He had always planned to farm. "He was the bluest of the whole bunch," Andy says.

Pat had little choice but to commute 50 miles away to a technical college in Norfolk, Nebraska, where he studied air conditioning and refrigeration and industrial maintenance. Julie was working as an elementary school teacher.

After graduating with straight-A grades, the best job he could find was on an assembly line at a car air-filter factory in nearby Yankton, South Dakota.

"I found out that I couldn't be someone else's moron for very long. It about drove me nuts but it was a job," he says.

Six months later, he found work in his field, with an air-conditioning contractor. He worked as a technician for about three and a half years.

Pat never got completely out of agriculture, though. After the farm sale, the bank that acquired the farm allowed Pat and Julie to rent the farmstead. Since they were buying a house in town, his parents stayed there. First with his father and then on his own, Pat "custom fed" pigs in the finishing building. He was paid a fee by the owner of the pigs, a local feed dealer. By then, the cycle of hog prices had turned profitable and there was no shortage of rural people willing to invest in hog production.

Until the fall of 1989, the bank rented the farm's cropland to different neighbors. That year, the bank wanted to sell. Fearing that the family might never have another chance to regain the farm, an aunt living in California decided to buy it. Andy's sister had been single until later in her life and had saved from a successful career as a nurse working on cruise ships. It was a break for Pat. His Aunt Liz rented him the land and he managed to farm on weekends. Pat traded his labor on farms of neighbors and relatives in exchange for borrowing their machinery.

The next year, an uncle, who was retired as the farm manager at Boys Town in Omaha, loaned Pat and Julie $16,000 to buy a small herd of cows. Pat agreed to pay the uncle the same interest he had paid to borrow the money in the first place. It was a five-year loan. The uncle, who has never given up his interest in farming, had approached Pat to offer the cattle loan.

"He and his wife purchased the original cow herd we started out with," Pat says. "I don't know if I would have been brave enough to go to a bank and say, 'I want to borrow money to buy some cows.' I don't know if we would have been able to do it."

But the following year, in 1991, Pat did muster the courage to take on some modest debt. He applied for a "limited resource" loan from Farmers Home Administration to buy half of the farm from his aunt. Farmers Home Administration, the lending arm of the U.S. Department of Agriculture, at that time was making a small number of low-interest loans to farmers with modest financial resources.

By then, Pat had been borrowing for five years from a bank for some expenses for his hog business, repaying the loans from proceeds from his custom finishing.

That helped Pat and Julie get the government loan for $124,000. "Custom feeding hogs gave us a little bit of leverage with Farmers Home Administration because they like to see somebody that has some experience, a little bit of a record," he says.

Tempered by his experiences from the 1970s and '80s, "when we expanded and went gung ho on buying the latest equipment," Pat has become very cautious about buying anything he doesn't need. He spent only $2,500 to build 20 pens in his barn for farrowing sows. He now farrows about 35 sows two times a year and finishes out his own pigs for market. His methods are competitive enough that his break-even price is about 35 cents a pound. Many farms must get about 40 cents to break even.

Pat has also given up growing crops and seeded the half of the farm that once grew corn, oats and a few acres of soybeans back to grass. In one of his first years of farming, "we went about 60 days without rain." Crop production seemed too risky, and too expensive. Abandoning it was a decision that he and Julie made together after taking a course in holistic management, an approach to business management that starts with setting clear, long-range goals and tries to reach those goals by looking at the farm as one system.

Using inexpensive electric fence, Pat has divided the farm into 26 paddocks of about 15 acres each. The farm supports about 150 cows and heifers. By rotating the cows from one paddock to another after a few days, Pat gets more forage production from his land. He buys no corn or supplemental feed for the cattle and puts up his own hay for the winter months. Pasture rotation has long been common in Europe and England and is hardly a new system. But farmers in New Zealand have improved the technique with electric fence. New Zealand dairy farmers can produce milk for about half the cost on U.S. farms, partly because of a milder climate but also because of their skillful management of pasture.

Pat and Julie still live mostly on the income from Julie's job as a remedial reading teacher at an elementary school and they're putting all of the farm's income back into the business. In their first three years of buying the farm, they have reduced their land debt from $124,000 to $97,000. They owe only $9,000 on the $16,000 cattle loan. The $2,500 loan to buy sows and build farrowing pens was paid off after the first year.

Pat admits that it would have been difficult to start out without the help of his aunt and without the Farmers Home loan. But he thinks that other young people with modest means can begin farming.

"Don't go out and borrow a bunch of money," the stocky young farmer advises with a wry smile.

As her two-year-old daughter, Angela, squirms on her lap, Julie adds that "you've got to have off-farm income. And don't think you have to have the biggest and the newest" farm machinery and equipment.

Not Alone

Moe Russell, a native of Monticello, Iowa, has worked 24 years for a farmer-owned network of lending institutions called the Farm Credit System. Today he's division president for branch lending for Farm Credit Services of the Midlands, an Omaha-based part of the Farm Credit System with $3.8 billion loaned to farmers and ranchers in Iowa, Nebraska, South Dakota and Wyoming.

Russell has known thousands of families like the Steffens, and he has seen his own association's lending practices grow much more conservative after the Farm Credit System required a federally-backed bond sale of up to $4 billion (which the Omaha bank repaid 11 years early).

When Russell was starting his career, farmers routinely started out in farming with as little as 10 percent down if they got a Farmers Home Administration loan. The Farm Credit System was nearly as liberal, financing up to 85 percent of a farm. "And they were the first ones to get into trouble," he recalls.

"Today, Farm Credit Services and other lenders have many excellent products [types of loans] to help farmers manage interest rate risks. The industry didn't have these in the late 1970s and early '80s and I think that contributed some to the economic hardships that were experienced by producers," Russell says. "For example, a dairyman in 1975 may have made a sound business decision in expanding his herd and building a milking parlor and loafing barn and feed storage facilities. But simply because that fixed investment was tied to a variable rate note, fixed costs went up when interest rates went up and put a lot of producers out of business."

At the time, most lenders were less concerned about how much money a farm made. Land values went up every year and if a farmer had pledged land as collateral for a loan, land inflation seemed like safe protection from many temporary losses. That was until the early 1980s, when so many farms were affected by high interest rates that too much land and machinery went on the market at once. A 50-year trend of rising farmland prices ended with what amounted to a crash in the land market. Midwest farmland dropped in value from June 1982 until February 1987 by about 65 to 72 percent, depending on the location, Russell recalls.

Falling land values meant that the collateral backing loans to farmers was suddenly worth less. Bank regulators began to pressure banks to call in the most-troubled loans. Some small country banks with a large portfolio of agricultural loans failed. When they were taken over by the Federal Deposit Insurance Corporation, the FDIC's liquidators took over farms, machinery and livestock with an efficiency that seemed draconian and heartless to many in rural America.

Land prices did bottom out in 1987, after the federal government pumped billions of dollars into the farm economy through crop subsidies and put up even more money for Farmers Home Administration guarantees of bank and Farm Credit System loans.

Today, Farm Credit Services rarely makes a loan for land or a capital improvement that requires less than 30 percent down, says Moe Russell. And, because farmers face so many risks—from weather, changes in commodity prices, and fluctuating interest rates and asset values—Russell's Association requires a comfortable level of working capital. It likes to see a farmer have enough working capital so that during a break-even year, the liquid position can absorb the principal payments, ideally up to three years.

"Farming is an unbelievably tough business and financing farmers is unbelievably tough," he says.

The Aftermath

Against that backdrop, it is no wonder that when the U.S. Department of Commerce conducted its twice-a-decade Census of Agriculture in 1987, it showed an exodus from America's commercial farms.

But you had to look carefully at the statistics. The drop in farm numbers was actually smaller in the "farm crisis" decade of the 1980s than in the boom years of the 1970s. In that earlier decade farmers who wanted to cash in on their life's work could easily find an urban doctor or lawyer to buy their land as a hedge against inflation and as a tax shelter. Or they could take the more common approach of just selling to a neighboring farmer. Farms expanded in size and farm numbers dropped.

But, in the 1980s, two unusual things happened.

First, older farmers seemed to stay in farming longer. Some who might have retired didn't want to sell their land in a depressed market, unless forced by a lender. Older farmers who had been conservative in the 1970s had money to buy land when it hit bargain basement prices in 1987. Others who had retired and financed the sale of their land through contracts with younger farmers wound up getting their land back when buyers could not make the payments.

Second, a real exodus took place among some age groups. Some middle-aged farm families with children who might succeed them quit, or discouraged their children from pursuing a farm career. Other, younger farmers who had recently borrowed to start farming or to expand their businesses were caught in the interest rate squeeze and forced out of business.

When the 1987 census was taken, the average age of the nation's farmers had reached 52. In 50 years, the percentage of farmers older than 55 had gone from 35 percent to 45 percent. The overall trend had not changed when the 1992 Census was taken, either. At that time the average age of farmers was up again, to 53.3 years.

The census also compiles data on each state. Iowa showed the effects of the farm crisis exodus clearly and is a good bellwether for trends in commercial family-operated farms. Although Iowa lags behind the diversified agriculture of California in total sales, in many ways it is the nation's most agricultural state. More of Iowa's surface area is planted to crops than any other state in the nation. In normal years it's the nation's top producer of corn and of hogs and often ranks first or second in soybeans. Like wheat farmers in Kansas, cotton farmers in Mississippi and rice farmers in Arkansas, Iowa's corn farmers also depend on federal farm programs for a significant share of their income.

In Iowa, the number of farmers under 25 fell by 55 percent between 1978 and 1987. In contrast, the number of farmers over the age of 75 increased by nearly 32 percent. Between 1959 and 1978 the trend had actually been just the opposite. Partly because of the boom of the 1970s, when increased grain exports to the Soviet Union and inflation made farming seem very attractive financially, the number of Iowa farm operators under 25 had grown, rising from 4,336 in 1959 to 6,339 in 1978. (That was a small percentage of the 105,000 farmers in Iowa in 1987, but the number of independent farmers under 25 has been small for decades, because farming on your own requires more capital than is available to most rural people under 25.)

"I've heard it said many times that we lost a generation of farmers," says Jeff Plagge, a banker in Waverly, Iowa, who has brought the loss of young farmers to the attention of the American Bankers Association.

The biggest changes took place in the worst farm debt crisis years between the Census of 1982 and 1987. The number of Iowa farmers under 25 fell by 49 percent, while the number of farmers over 65 increased by 20 percent. Nationally, the loss of young farmers was nearly as great—a 42 percent drop.

Some analysts have pointed out that the steep decline in numbers of young farmers didn't really represent that many farms going out of business. Instead, many young people just didn't start out in farming in the first place.

But other relatively young age groups that probably represented active farmers did fall off, too, during the 1980s. The number of Iowa's farmers between 25 and 34 fell by 14 percent. And the number of farmers between the ages of 45 to 54—families most likely to have children in college, technical schools or about to enter a career—fell by 17 percent. Just like Andy and Rita Steffen of Fordyce, Nebraska, older families forced to leave agriculture could not have helped their children much, even if those children wanted to farm.

By 1992, when the next Census of Agriculture was taken, the trend in fewer young people entering farming was continuing, but there were some signs that it may have been leveling off a bit. The number of farmers under 25 declined by 20 percent, less than half as fast as in the previous five years.

But the dropout rate of the next oldest group of farmers, between 25 and 34, was actually worse than in the previous five years—25 percent. A more seasoned 35 to 44 age group increased just slightly.

Overall, though, the average age of Iowa farmers was still creeping upward, from 47.6 in 1982 to 49.3 in 1987 to 50.3 in 1992. The group over 65 was growing, too, with those over 70 growing dramatically, by 19 percent.

"There is a dwindling pool of people who can come up with the money to buy land" in rural Iowa, says Paul Lasley, a rural sociologist who has done some of the best analysis of these demographic trends and what they portend for the future of the state's farming and rural communities.

Another reason that the older farmers haven't retired is that land seemed to be a better investment than selling a farm and putting the money in the bank, Lasley says. Bank interest rates for savings deposits were unusually low in the early 1990s.

Lasley also thinks that the idea that most farm families can live entirely off their farm income may be a historical aberration. Even in the nation's colonial period, when a large share of the population lived on farms, "they had small farms and they had off-farm income," Lasley says. "They were blacksmiths, or farriers or carpenters or bricklayers, miners, lumbermen."

But Lasley doesn't minimize the significance of the loss of the state's farmers. "Iowa stands at the crossroads in deciding the future farm structure of the state," Lasley wrote in an analysis of the Agricultural Census. "The debate over the structure of agriculture has a long history dating back to the early land disposal acts and settlement period. ... It appears Iowa agriculture is poised for a substantial shift in the next couple of decades."

At the national level, the 1992 Census of Agriculture showed trends similar to those in Iowa, with the decline in the very youngest, under 25, age group slowing down but the next oldest group actually losing members even faster in the previous five years. Every age group grew smaller, except for the number of farmers 70 years old or older. That group increased by 13 percent.

Whatever happens, Pat and Julie Steffen are doing their best to prepare to survive the changes in agriculture. During a break from talking about his first years in farming, Pat and his two sons, six-year-old Robert and five-year-old Tom, walk up the hill from their house to move the cattle herd from one pasture to the next.

In three days, the cows have mowed off the alfalfa and orchard grass in their six-acre paddock. They bellow and move toward him as Pat and his sons approach the fence. Pat carefully lifts each boy over the electrified wire into the grazed pasture and they walk toward the top of the hill, where Pat disconnects a hot wire separating the cattle from a field of tall, lush forage. When he drops the wire, the cows and their calves rush into the next field.

Pat lets Robert and Tom take a turn at reeling in the spool of disconnected electric wire while he points out that the pasture still has terraces. This was a field where his father had grown oats and alfalfa in a crop rotation with corn. The terraces helped stop erosion, he says, but the soil was light and sandy in spots. He kicks up a patch of light brown soil in the grazed pasture now empty of cattle.

To get farm program payments from the government, his father had to plant corn each year in order to remain eligible. Although he, too, used farm price support programs when he started out, Pat wants as little to do with government farm programs as possible.

"I think, for a lot of people, it's like alcohol. They get hooked on it and they stay on it," he says.

Later, Pat and his father show a visitor the infamous hog-finishing building. There are two large holes cut into the south side framed with two-by-four lumber. Pat has tried to increase the natural ventilation to cut the electric bill for the fans.

"This barn setup cost $65,000 and now he's tearing it apart," Andy says.

Pat adds that he plans to open up the entire south side and replace the wall with a heavy plastic curtain that can be raised and lowered as needed to keep the hogs comfortable. He plans to build it from several discarded giant bags used to store ensilage.

"It won't look quite as good but it's free," he says, comparing it to ready-made curtains he could buy. "A little inconvenience is easier to put up with than no money."

As we'll see, that may be the motto of a whole generation of farmers.

CHAPTER 2

The Treadmill of Technology: Beginners Have to Run Faster to Catch Up

Glenn Holderread, an Oklahoma dairyman who was among the first to start using the controversial hormone bST to increase milk production. Usually, such innovations attract little attention outside of agriculture. Photo by Chester Peterson, Jr., courtesy of *Successful Farming*.

In the spring of 1994 milk prices were falling. They fall every spring as dairy cows are turned in to fresh pasture and milk production rises. So falling prices alone weren't anything remarkable.

But that year something new was adding to a glut that forced dairy processors in California and Texas to dump milk by early May. According to a story in the *Milwaukee Sentinel*, it was a new and controversial product of genetic engineering—recombinant bovine somatotropin (rbST), also known as recombinant bovine growth hormone (rbGH).

The synthetic hormone, similar to one that occurs naturally in cows, is made by genetically altered bacteria. Dairy farmers buy it to inject into cows during part of the lactation. If the cows are high milk producers that are well fed and well cared for, injections of extra hormone can boost milk production by an average of some 12 percent per cow.

After a long, bitter struggle that pitted consumer groups and small-farm advocates against companies that developed rbST, the Food and Drug Administration approved the hormone's use in November 1993. Monsanto Company began selling the hormone as Posilac in February 1994.

It seemed to be the beginning of a what small dairy farmers feared. Not only would rbST boost milk production, but the FDA's approval of rbST didn't reassure consumer groups, including the Consumers Union, of the hormone's safety. If the milk supply rose and consumer confidence in milk fell, prices would drop. Big, high-volume dairy farms of the south and west were expected to be the survivors in this process. Already in the first three months of 1994, milk volume in California was up 8 percent over the previous year. Milk production in Wisconsin was down. Prices on the wholesale cheese market were falling.

"Farmers who use the growth hormone are only getting a short-term fix. By fall they will be losing money hand over fist as prices drop," Lee Light, a Marshfield, Vermont, dairy farmer said in a statement released by the advocacy group, Rural Vermont. "Meanwhile, they are poisoning milk prices for the 90 or 95 percent of the dairy farmers who are not using BGH [rbST]. Today, we are asking the entire dairy industry—farmers, dairy co-ops, milk processors, and grocery stores—to 'just say no' to BGH."

The bitter struggle that greeted the introduction of rbST to the dairy industry was an unusually strong reaction to new agricultural technology. And it was one of the few changes in farm technology that caught so much attention from the general public. But it wasn't entirely unique.

American agriculture's history is one of technological change, from the cotton gin of the 18th century, to steel plows, grain drills and threshing machines in the 19th century, to a dizzying pace of innovation in the 20th century—with hybrid corn, gasoline tractors, chemical fertilizer, computers and biotechnology. The nation's farmyards are littered with relics of technology considered outmoded—small tractors, corn pickers and other useful inventions muscled aside by higher-volume machines.

The farm debt crisis of the 1980s that hit the Steffen family of Nebraska is the force that many Americans will remember driving farmers from the land. Farm foreclosures inspired movies, Farm Aid concerts by country singer Willie Nelson, and congressional hearings. But since about 1935, when farm numbers peaked at 6.8 million, technology has been the quiet reaper of family farms. And, because technology has increased farm size and made entry into farming more costly, the overall trend has inevitably limited the opportunities for beginning farmers. Yet, as we shall see, not all types of technology are barriers to be surmounted. Some technologies offer innovative young farmers a start in this race for survival.

In the century drawing to a close, technology has made rural life easier and perhaps more pleasant. For consumers it made food cheap, if sometimes dull and bland. But it also reduced the need for labor on farms. Tractors, widely adopted by farmers in the 1950s, "decreased the need for hired labor and tenants for most aspects of crop production," writes Iowa State University agricultural historian R. Douglas Hurt in his book, *American Agriculture: A Brief History.*

Hybrid corn introduced in the 1930s and cheap chemical fertilizer adopted by farmers after World War II began a "green revolution" which also dramatically increased crop yields worldwide. But these higher yields tended to lower the unit price for crops, so farmers often had to expand their acreage to maintain or increase income. The decline in farm numbers accelerated in the years after World War II.

By 1990, farm consolidation had cut commercial farm numbers to just over 2 million. And the number of Americans who were farmers had fallen from a fourth of the population in the Great Depression to a barely noticeable 2 percent.

Those who understand agriculture know that farmers don't always greet change the way some small- and medium-sized dairy farmers have fought the introduction of rbST.

Marty Strange, an economist and founder of the Center for Rural Affairs, a Nebraska advocacy group, has criticized the way the agricultural establishment has replaced skill and husbandry with technology that has negative effects on people and the environment. Yet, when Strange defines his ideal of the family farm in his book, *Family Farming: A New Economic Vision*, he doesn't portray farmers as nostalgic or as rebelling against all innovations. Strange's admittedly idealized definition of family farmers includes those who are, as he puts it, "technologically progressive."

> Family farming is innovative, using technology to reduce costs and to lighten the load on human beings, but its objective in employing new technology is to enhance the work of the family, not to eliminate work for people. There is, of course, an internal contradiction here. New technology that enhances one family's labor tends to displace another's in the long run, but it is not this consequence which motivates adoption of the technology. Instead, such a consequence is viewed with alarm, if indeed, it is understood at all.

Many thoughtful farmers do understand the effects of technology. They know that nearly every year, unless there is a severe drought or natural disaster, the nation will produce more bushels of wheat, corn and soybeans, thanks in part to more productive new varieties. It's that knowledge that drove corn growers in the Midwest to write the White House during the busy planting season of 1994 when they thought that the federal government might abandon a commitment to require that fuel ethanol be added to gasoline under new rules for the Clean Air Act. The rules would create a growth market for corn used to make that ethanol. It's why wheat farmers supported export subsidies for wheat in the 1980s. And it's why many hog and cattle producers favored liberalized trade rules and the North American Free Trade Agreement. A slowly growing population in the United States will eat only so much meat, but the growing middle classes of Asia and Latin America seemed like potential markets.

Other farmers, who favor more government intervention, would like to see programs to cut production or manage the supply of farm products. But, unless such programs could be sold to an increasingly urban Congress as "green payments" for saving the environment, such intervention seemed unlikely in the conservative mood of Washington in the 1990s.

The general effects of technology on agriculture are so well known that technological change was described as early as 1958 by agricultural economist Willard Cochrane as a treadmill. In 1965 he explained this to the general public:

> The innovators reap the gains of technological advance during the early phases of adoption, but after the improved technology has become industry wide, the gains to the innovators and all other farmers are eroded away either through falling product prices or rising land values, or a combination of the two, and in the long run the specific income gains to farmers are wiped out and farms are back where they started— in a no profit position. In this sense, technological advance puts farms on a treadmill.

David Campbell, a political scientist at the University of California in Davis, has found a lot of scholarly studies and evidence that suggest that rbST, the synthetic dairy hormone made with biotechnology, will be another treadmill.

Campbell looked at the potential effects of rbST on dairy farms and rural communities for the University of California's Sustainable Agriculture Research and Education Program. Campbell and other scholars compared rbST with another technology less likely to put dairy farmers out of business—rotational grazing. That's the same system that beginning farmer Pat Steffen uses for his beef cow herd in Nebraska when he divides pastures into small paddocks with electric fence, moving the herd in a rotation from one paddock to another. The comparison of the two systems was published in 1993 as the book, *The Dairy Debate: Consequences of Bovine Growth Hormone and Rotational Grazing Technologies.*

Campbell found eight studies that suggested this likely outcome of the drama that began with falling milk prices in the spring of 1994:

- The new synthetic hormone would increase milk production per cow by 10 to 15 percent.
- The hormone would be widely used, by about 40 percent of dairy farms in the Pacific region and by 25 percent of farms in the Corn Belt after 5 years.
- Larger farms will start using rbST soonest and will benefit the most, because the farms will generally have better management needed to use the hormone and because they own more higher producing cows, which respond with better milk production from rbST than lower producing cows on smaller farms. Campbell concluded:

> While none of these eight studies claims that bGH [rbST] alone will be responsible for rural community decline, taken together they provide little or no evidence that the product will aid dairy-dependent communities, and fairly strong evidence that widespread bGH adoption will lead to substantial harm in certain local areas.

Like most farm policy debates, the one over dairy technology has focused on the demise or growth of *existing* dairy farms, not what this technology will do for *beginning* farmers who'd like to start in dairy production. But Campbell's analysis is hardly encouraging:

> A key issue is whether the new technology is scale neutral or if, instead, it imposes capital, labor or management costs that make it more likely to be adopted by the larger and more economically secure producers. Most major advances in agricultural technology have not been scale neutral, and resulting adoption patterns frequently have favored larger farms.

Many beginning farmers are hardly "larger and more economically secure."

Yet, while the debate over rbST raged in newspapers and the popular press, some young farmers who were definitely not economically secure, started in dairy production just when the industry seemed on the verge of a huge technological leap that would leave them behind.

Two of those young farmers were Steve Hopkins and Sara Andreasen who started farming near Decorah, Iowa—in the scenic northeast corner of the state that is an extension of the Wisconsin and Minnesota dairy belt. Steve and Sara's systematic approach to finding land and farming opportunities will be described later in this book. But their approach to dairy production has a lot in common with Pat Steffen's frugality. Steve and Sara are using rotational grazing to help keep their production costs much lower than is typical on established conventional dairies.

Their resources at the start of their career in 1992 were indeed modest. Neither Steve nor Sara had parents who farmed. After both worked at other jobs for several years—Steve for the Extension Service and Sara as a school social worker—they had managed to save $20,000 to start farming. They bought $1,500 of electric fencing and a manure spreader and rented everything else, 15 cows and 20 acres of pasture, which was more than enough using rotational grazing techniques.

As Steve explained at the first national conference for beginning farmers and ranchers in Omaha, Nebraska, in 1994, "It's often said that farming is going high tech. I propose that high-tech farming is one way and another way is what I call 'high technique.' When I think of high tech, I think of an expensive technology that's usually manufactured by a scientist somewhere and you have to pay a lot of money for it and you're oftentimes stuck with buying a certain product. ... High technique, I consider a type of technique that requires skill. It's sophisticated and yet it's low cost. That's what I consider our intensive grazing techniques to be, a high technique type of operation. And that's what we're aiming to do on our farm, to keep our costs low, to use our own skills and not have to be sending checks into a laboratory for gee whiz science type techniques." Steve and Sara are substituting their time and labor (which involves moving their cows into a new paddock twice a day, after each milking) for the cost of buying a high-energy feed and for using rbST.

They're using their knowledge of rotational grazing technology, which Steve learned about when he worked for the Extension Service and on a trip to New Zealand—the nation which has refined the art and science of rotational grazing.

Although Steve calls rotational grazing a technique, because it demands a lot of skill to manage both the grass growth and the animal nutrition in a rotational grazing system, it's still a type of technology, even if it doesn't require a lot of capital.

Steve's feeling that rotational grazing can be competitive with the "high tech" approach, especially using rbST, has impressive scholarly support.

Other experts who contributed to the book, *The Dairy Debate*—University of California agricultural economist L.J. "Bees" Butler and graduate student Gerry Cohn—compared using rbST and rotational grazing and found rotational grazing competitive. They analyzed the expected costs and returns from both systems on a farm with 100 cows. Before making any changes in production, the farm had annual net revenue of $399 per cow. After using rbST, net revenue increased to $444 per cow. But, if the same farm switched to using rotational grazing, net revenue also increased to $444. Butler and Cohn concluded:

> As would be expected, total revenues under bGH [rbST] increase due to increased milk production. BGH also increases feed costs, milk hauling costs, labor costs and veterinary/nutritionist costs, thus increasing total costs. Under rotational grazing, on the other hand, total revenues are reduced due to a decrease in milk yield [by 5 percent], but the savings in feed costs, fuel and electricity, labor and machinery lower total costs despite offsetting increases in fertilizer and fencing costs.

Unlike the hormone, which causes a dairy cow to produce more milk—and consume more purchased resources—rotational grazing saves costs by increasing forage production, almost doubling it in some cases. When livestock are confined to a small pasture for a short time, they eat all of the plants more uniformly, in effect, mowing off the pasture. Then, if the plants are given enough time to recover, they remain in a productive, vegetative stage longer before going to seed. Pasture rotation is nothing new, but modern electronics which deliver short bursts of high voltage for electric fencing have offered an inexpensive way to divide up pastures.

The study by Butler and Cohn also looked at how the two technologies would affect dairy farms at a national and regional level. Nationally, it suggests that if milk prices fall, "dairy farmers are better off (more profitable) under rotational grazing than under bGH."

If prices rise—which seemed unlikely in 1994—farmers using rbST would be better off because they would have higher output of milk. The two economists pointed out that their study was a relatively simple analysis of some reasonable hypothetical budgets for the two systems. And, they said, it's possible that some farms will adopt both rotational grazing and rbST.

But the clear implication of the study, and of the entire book, *The Dairy Debate*, is that some technologies are more likely to speed up the treadmill of technology than others. And some, like rotational grazing (also called controlled grazing and management-intensive grazing) are more farmer-friendly than others. Often, this technology has been developed by farmers rather than university or corporate researchers.

Another example of less costly technology is "ridge tillage," a method of crop production that helps lower soil erosion. With this system, farmers build up a ridge while cultivating weeds during the summer, then use special equipment to plant into that ridge as a raised seedbed the following spring. Because farmers can "band" weed-killing herbicides and fertilizers only in the seedbed instead of over the entire field, ridge tillage lowers the investment they need to make in purchased farm chemicals. Like rotational grazing, it requires knowledge and skill to make it work as well as no tillage, another soil-saving technique.

Unfortunately, as state agricultural colleges have faced the same budget pressures facing other government institutions in the 1980s and 1990s, many have become more dependent on private grants from companies that are developing the "high-tech" agricultural innovations that Steve Hopkins and other beginning farmers find too expensive to be practical. Independent, basic research that might benefit young farmers has become a scarce academic commodity.

There are some bright spots, though. Several state universities, including some in Iowa, Nebraska, Minnesota and California, have special research projects on sustainable agriculture. True sustainable agriculture is more than using environmentally sound methods that our natural resources can sustain for generations. To many, sustainable agriculture also means methods that are profitable and accessible for the average farmer, including the beginning farmer. They are methods that will sustain rural communities and independent, family ownership of farms. The universities that have these programs on sustainable agriculture will have something to offer young farmers.

One of the brighter lights showing the way to use "technique instead of high-tech" is Jim Gerrish, a University of Missouri research agronomist based at the University's Forage Systems Research Center near Linneus, Missouri. Since 1990, Gerrish has organized several three-day grazing workshops for farmers during the growing season. It's taught by University of Missouri agronomists and animal scientists, representatives of the U.S. Soil Conservation Service, and by farmers who use controlled grazing.[1] The "grazing school" has earned a national reputation that draws farmers from across the United States.

---

[1] For information, write Gerrish at MU FSRC, Route 1, Box 80, Linneus, MO 64653.

CHAPTER 3

Rural Boardwalk: How Shared Monopoly Threatens Young Farmers

Roy Henry, a Kansas hog producer who organized an informal trucking co-op that has helped increase market prices for him and his neighbors. Henry served as president of the Kansas Pork Producers Association in 1995. Photo by Ed Lallo, courtesy of *Successful Farming.*

To many Americans, the word *monopoly* means little more than a board game. Real monopoly doesn't seem to affect their lives. They may, in fact, benefit from the economic clout of their employer in the marketplace if they work for a large company. To farmers, and the minority of Americans who run their own businesses, the effects of restricted competition are as real and worrisome as natural disasters and recessions. If there is less competition among those who sell products to farmers or who buy farmers' crops and livestock, farmers will earn less. For a beginning farmer with modest income and perhaps some debts, the effect of this potential squeeze can be even more crucial.

If three or four companies control more than 50 percent of a market together, they often develop some degree of "shared monopoly power" that affects prices. The higher the level of concentration, the closer the shared monopoly power comes to that of a single monopolist or a cartel. The technical names for concentrated markets of this type are *oligopoly* for sellers

21

and *oligopsony* for buyers. When concentrated industries also have strong brands, as in the breakfast cereal and soft drink industries, their monopoly power is even greater.

Farmers may not know all of the technical jargon that economists use. But in an era when the general public seems to have lost interest in real monopoly, farmers are keenly aware of the trends.

They have a heritage of interest in competition. Many farmer cooperatives grew out of the populist movement that was a reaction to monopoly power. The nation's first effort to restrict anticompetitive practices, the Sherman Antitrust Act of 1890, had its roots in agriculture. The law was passed to correct lack of competition in the meatpacking industry. It later was used to break up powerful business trusts—agreements that restricted competition—in oil and tobacco.

The merger mania and consolidation of big business that took place nearly a century later in the 1980s directly affected farmers. By the late 1980s and early 1990s, rural sociologist William Heffernan and his University of Missouri colleagues had turned tracking the mergers into a time-consuming project.

They found that the four biggest meatpackers controlled 72 percent of the beef slaughter and 45 percent of the pork slaughter. Only 20 big commercial feedlots produced half of the nation's fed beef. The top four soybean crushers and top four wet-corn millers controlled about three-fourths of those markets.

At a time when profit margins for grain farmers were growing thinner and when some farmers struggled with the worst financial crisis since the 1930s, food processing profits seemed unjust. *Forbes* magazine published the return on equity for General Mills at nearly 49 percent in 1993. ConAgra's return was better than 20 percent. The giant beef and pork packer, IBP, had a more modest 6 percent return.

It's important to point out that most farmers hardly balk at some long-established market power. Deere and Company has a large share of the market for some types of farm machinery. Pioneer Hi-Bred International sells about 40 percent of the seed corn in the United States. Yet farmers are willing to pay higher prices for John Deere tractors and Pioneer seed. A reputation for quality products—not mergers that might be challenged under antitrust laws—seems to have gained those companies big shares of the market. Nor did it seem to bother many grain farmers that ADM (Archer Daniels Midland) made 60 percent of the fuel ethanol produced in the United States from corn. After all, ADM was creating new markets for corn in a grain market long dominated by a half-dozen giant international companies.

For young farmers, a more worrisome change in the marketplace is the growing domination of livestock slaughter by a few companies. Cattle and hog production has been generally profitable. In the hog business, the long-term profit as a percent of costs has been about 15 percent. And for young farmers in the Midwest, hog production was often the least expensive and most profitable way to start out. A fifth of the nation's farms producing feeder pigs were run by farmers between the ages of 25 and 34, according to the 1987 Census of Agriculture.

Because cattle and hogs usually were profitable for producers in the 1980s and 1990s and packers sometimes competed fiercely for a small supply, prices seemed competitive on the surface. But farmers in some areas were losing out. Bruce Marion, an agricultural economist at the University of Wisconsin in Madison, is a national authority on the effects of decreasing competition in the meatpacking industry. In a report he published in 1993, Marion documented the decline in competition among food manufacturing industries and showed how shared monopoly by meatpackers was affecting prices paid to cattle and hog farmers. In *Status of Wisconsin Farming, 1993*, Marion wrote: "The 1980s witnessed a great surge in merger activities, including more large mergers than at any time in our history." He found a sharp rise in "concentration"—shared control of the market by the top four companies—in several food industries that don't sell branded products, including meatpacking:

> Mergers were a major cause of increased concentration in most of these industries. Several of these mergers would have been challenged by the antitrust agencies under previous [presidential] administrations. ... National concentration of fed steer and heifer slaughter increased from 26 percent for the largest four packers in 1972 to 30 percent in 1979. Four-firm concentration then rose sharply over the following seven years to 55 percent by 1986. ... As a result of three large acquisitions by ConAgra ... and Excel's acquisition of Sterling Beef, all in 1987, four-firm concentration increased to 68 percent by the end of 1987 and hit 72 percent in 1990. ... This rate of concentration increase is unprecedented. There is no parallel in any other industry. The industry is now dominated by three large companies, IBP, ConAgra and Excel [owned by Cargill], which collectively slaughter nearly two-thirds of all steers and heifers in the U.S.

Those are national statistics. The number of packers buying cattle varies a lot from one region of the country to another. By looking at price trends in different regions, Marion showed how much the decline in competition costs cattle producers:

> When all else was held constant, a region in which the top four packers slaughtered 50 percent of the fed beef had prices that were roughly $1.00 per hundredweight higher than a region in which the top four packers slaughtered 90 percent of the fed cattle.

At the end of the 1980s, the hog-packing industry remained more competitive than the fed beef-packing industry, but it, too, was consolidating. In a decade the top four hog slaughtering companies increased their share of the market from about a third to more than 40 percent by 1990. "IBP and ConAgra both entered hog slaughtering in the early 1980s and now rank number one and two," Marion wrote.

A graduate student working with Marion, Charles Heyneman, studied regional markets for hogs and also found a price-depressing effect where fewer companies slaughter hogs:

> All else the same, hog prices during 1977-89 were roughly $1.40 [per hundredweight] higher in regions where the top four packers slaughtered 50 percent of the market hogs than in regions where they accounted for 90 percent of the slaughter.

Over several years those differences can amount to significant income loss even for smaller feedlots run by farmers, let alone the large commercial feedlots in the High Plains of Texas, Oklahoma, Kansas and Nebraska. (But seasonal variation in prices in one year in the early 1990s could be $10 per hundredweight or more. So, when prices were good, Marion's work was easy for farmers to ignore.)

Larry Anton, a LaPorte City, Iowa, farmer who, with his brothers, feeds out 1,500 head of cattle a year has long had a strong interest in getting farmers to work together to improve their markets. Anton headed the marketing committee for the Iowa Cattlemen's Association and he sells his cattle through Interstate Producers Livestock Association, a marketing group originally formed by Farm Bureau that charges a small commission for negotiating the best prices from packers for its farmer clients. "We've always felt that was worth at least a dollar [per hundredweight]" says Anton of the higher prices the brothers receive. The marketing group seemed even more important in recent years as Anton saw small, independent packing plants in his area close. By the mid-1990s he was down to three packers, one of them quite small, that would be willing to send a buyer to his feedlot if he tried to negotiate prices on his own, without going through Interstate Producers Livestock Association.

Yet, Anton admits that it's been a hard sell to get farmers to work through marketing groups or to drum up interest in packer concentration among the leaders of livestock organizations.

That was especially true when prices were good. But in April and May of 1994, fed cattle prices suffered one of the worst crashes in years after the beef supply in feedlots reached the highest level in more than a decade. Anton saw the value of the cattle in his own feedlot fall from $77 per hundredweight to under $65. Some cattle feeders were losing $100 to $150 on each steer or heifer they sold.

"A break like this gets people more interested in marketing than when things are going along well," Anton says.

Not many young farmers start out in cattle feeding on their own because it can take a lot of borrowed money to buy cattle to fill a feedlot. But Anton had heard of one young man who had just started feeding cattle and was caught with 700 head in his lot during the crash. Anton's own son had left the family farming corporation not long before the crash and, although Anton felt badly that it hadn't worked out, he was relieved that his son had missed the disastrous downturn.

Anton was also one of eight members of the beef industry in four Midwestern states to serve on a "Task Force on Competition and the Livestock Market," that was organized by the Center for Rural Affairs in Walthill, Nebraska. The task force reviewed the condition of the beef packing industry and sought the advice of leading economists, including Bruce Marion, who had followed issues involving competition. In 1990 the group issued a report that concluded, as many producers had suspected for some time, that "competitive conditions in the cattle market are now inadequate to meet the requirements of a free market."

Not only did fewer packers buy more of the beef supply, there was ample evidence of other practices that reduce competition. Packers were contracting in advance with large feedlots to purchase an increasing amount of their supply directly. Such contracts created a captive market that enhanced packer power over prices. Only a small percentage of fed cattle are sold on the open market at livestock auctions and stockyards, making government reports of prices less reliable. Nor are all of the direct sales between producers and packer buyers reported. There was evidence that some packers had, at times, been able to manipulate prices in the way they used the futures market.

The task force recommended several changes in government action, including more vigorous enforcement of existing federal antitrust laws. It asked that enforcement of those laws be switched from the U.S. Department of Agriculture's Packers and Stockyards Administration to the Justice Department, saying that P&SA's record of enforcement was weak. It asked that packers be banned from holding more than 15 percent of the fed cattle supply "captive" through packers' own feedlots or through contracts with feedlots. It asked that Congress set up a mandatory price reporting system. And it asked that regulation of commodity futures trading be switched from the Commodity Futures Trading Commission to the Securities and Exchange Commission, which regulates the stock market and financial reporting by publicly held corporations.

By the mid-1990s, two years into the administration of President Bill Clinton, members of the task force saw little evidence that the new administration cared any more about competition in the livestock industry than Clinton's Republican predecessors, former presidents Ronald Reagan and George Bush.

"I guess I'm more concerned now than I was then about fair trade," says Dave Williams, a Villisca, Iowa, hog and cattle producer who also raises 300 acres of row crops in this southwest Iowa community. "I don't see the Justice Department or Packers and Stockyards increasing scrutiny. It seems like the IBPs, ConAgras and Excels have a bigger market share."

Williams, who farms with one son and who has helped two other sons get started in farming, is also worried about another significant change in the hog industry—the shift toward more vertical integration. That's another form of economic organization that can reduce competition. In the 1980s, the term, vertical integration, became a controversial household word on Midwestern hog farms. Vertical integration means that one company either owns or controls all of the stages of production. That was hardly anything unusual in much of the economy in the late 20th century. Oil companies owned their own wells, refineries, distribution systems and retail gas stations. Integration was still less common in agriculture. Independent farmers and ranchers raised calves and pigs. They might grow all of their own feed, or might buy some of it. Calves would be sold to other farmers or feedlots to finish to market weight. Feeder pigs would either be sold to other farmers to finish, or they would be finished to market weight on the same "farrow-to-finish" business on one farm.

The poultry industry had changed in the 1950s and 1960s to become vertically integrated, mainly through contracting. Independent farmers no longer raised a significant share of broilers. Instead, feed companies, food processing companies and others operated hatcheries that controlled breeding. The companies owned the chicks and contracted with farmers to feed them to market weight. The companies supplied the feed to their own specifications. The farmers were paid a fee for producing the broilers in their own buildings, but the buildings, too, had to be made to company specifications. The company also owned its own slaughter and processing plants. And by the 1980s, companies such as Tyson Foods, ConAgra and others were selling their own branded products in supermarkets. Independent poultry farmers had long since been locked out of that marketing system. They were no more likely to deliver a pickup truckload of chickens to a company processing plant than an independent Texas wildcat oil driller was likely to deliver a barrel of oil personally to an oil company refinery.

Vertical integration in the poultry industry caught the attention of the astute agricultural economist Harold Breimyer, an Ohio farm boy who helped write many New Deal farm programs and who spent his retirement years publishing an insightful newsletter out of the University of Missouri agricultural economics department. In the 1960s Breimyer wrote about the changes that had taken place in broiler production in the South and Mid-Atlantic states. One Delaware farmer who was a contract producer of broilers told Breimyer, "They say we are serfs, but I don't think we are." Breimyer thought it significant that the word, serf, had even entered the vocabulary of American farmers, whose ancestors had left Europe to escape the vestiges of serfdom.

In the Midwest of Dave Williams in the late 20th century, the hog industry was changing rapidly. Not far to the south of his farm, in northern Missouri, Continental Grain Company planned to build two hog operations that would farrow 20,000 sows. Tyson Foods, one of the nation's largest hog producers as well as a leading broiler company, owned a packing plant. In Iowa, there were reports of farmers getting letters from packing companies who refused to buy their hogs anymore. That was because the plants wanted hogs that produced less fat and leaner meat. But big producers who could deliver large quantities of hogs to packing plants were also reportedly being paid premiums of $2 to $5 per hundredweight. And in states as far away as Utah, big hog producing companies were expanding production.

By 1993, 57 very large swine producers sold 13 percent of all of the hogs in the United States, according to a survey by University of Missouri agricultural economist V. James Rhodes. That included seven "mega" producers who sold more than 500,000 hogs a year and 50 "super" producers marketing more than 50,000. Rhodes found 15 vertically integrated firms, but was surprised that they weren't growing as fast as other large firms. Nearly all big firms contracted production, though.

A year later, in the October 1994 issue of *Successful Farming* magazine, livestock editor Betsy Freese showed that the trend was accelerating.
By 1995, the nation's 30 largest pork producing businesses would sell one-
fourth of the nation's hogs, the magazine predicted.

In his newsletter, Breimyer said that some economists attribute these
changes in the industry to consumers' growing preference for lean, consistent quality pork and that the big producers were able to supply it:

> In all the rhetoric almost no one mentions what this column regards
> as the strongest driving force of all [in the changes in the hog industry].
> It's the struggle for, and exercise of, market power. Power, once ac-
> quired, tends to snowball; the bigs keep getting bigger.

The greatest risk to independent producers, Breimyer adds, is that vertical integration kills off open markets. Even if independent producers are
just as efficient as bigger producers, they will have difficulty surviving if
they don't have access to competitive markets.

Evidence of growing market power was reported in the March 1995
issue of *Successful Farming*. It's special report, "Market Power," showed
that some of the nation's largest hog producing companies had long-term
relationships with packers or owned packing plants. In some cases, large
producers negotiated risk-sharing agreements with packers that leveled out
prices packers paid to the producers. Two of the nation's largest packing
companies, Excel and Monfort Pork, indicated that they had a preference
to offer such contracts mainly to pork producers who had annual production of at least 10,000 head of market hogs.

Farmer Dave Williams was pessimistic about all of this. "I'm afraid
the little guy is going to further get squeezed because there are so many
new hog facilities getting built," said Williams, whose own farrow-to-finish operation with one son markets about 2,000 hogs annually. Another son
sells about 1,600 hogs a year. A third son is starting up a 250-sow business
that will sell feeder pigs.

For several years Williams served on the board of directors of Farmland Industries of Kansas City, one of the nation's largest regional cooperatives. Farmland, too, has begun its own vertical integration. It buys hogs
from farmers on the open market at its packing plants in Denison, Iowa;
Crete, Nebraska; and Monmouth, Illinois. But it also has begun to contract
with farmers to raise its own hogs for Farmland packing plants. Williams
says the contracting venture was done mainly to insure that the plants
would have a steady supply of high quality, lean hogs. But at first, Farm-

land's move into hog production stirred up resentment among farmers in Iowa, Minnesota and elsewhere in its trade territory. The farmers viewed Farmland as a competitor with independent producers.

Williams concedes that Farmland handled relations with its own farmer members poorly, but he laments the fact that few farmers seem interested in controlling the vertical integration process through co-ops that they (or their local co-op elevators) own. Not only does a co-op packer compete with the big companies that dominate the industry, but farmers potentially share in the co-op's profits by receiving patronage dividends (although many large co-ops are notoriously slow about paying out those dividends).

"The Scandinavian countries market about 95 percent of their hogs through the cooperative marketing system," Williams says.

Chris Hurt, an agricultural economist at Purdue University who has studied the pork industry's changing structure, says that some independent producers are already forming co-ops to compete with the "integrators" and he expects to see more. But he thinks they will be smaller, home-grown entities. "Our guess is it will not be existing co-ops. They are too monolithic and cannot adjust to the needs of the producers," he says.

Near Clay Center, Kansas, an informal, independent marketing group of 15 hog farmers is doing exactly what Hurt has predicted—and, ironically, is selling hogs to the Farmland plant about 100 miles away in Crete, Nebraska. Such marketing groups may offer a small ray of hope to beginning hog farmers starting out under the cloud of shared monopoly and vertical integration.

Steve Luthi, 34, who raises 800 acres of wheat and 500 acres of milo and soybeans with his dad, bought a half-interest in his father's farrow-to-finish hog business in 1980. In 1982 he bought a 15-acre farmstead with a hog farrowing building of his own. Today, the father-son operation has 225 sows and markets about 3,600 hogs a year. Nearly all of them are sold through the informal group organized by breeding-hog producer Roy Henry.

About eight years ago, when Henry's farm became large enough to consider delivering hogs directly to Crete instead of to a local "buying station"—a pickup point for the packer—Henry still had only enough hogs to fill half of a semitrailer for a weekly run to Crete. So he recruited neighbors to fill the truck with their hogs. The Luthis were early converts.

"We'd blow a whole morning" delivering hogs to the buying station, Steve says. The truck ordered by Henry stops by the Luthi farm and takes only a few minutes to load.

Such marketing groups also have the potential to pick up the $2 to $5 premiums some packers will pay to larger hog suppliers in return for consistent delivery. Henry is reluctant to talk in great detail about the prices his group gets, but he says "it's worth a strong base bid to do this."

The group has other advantages, too, he says. The packer pays each member for his hogs according to how much lean meat dresses out of each hog carcass. Members can compare each other's "kill sheets"—a printout showing payment on each hog—in order to see how efficiently their own hog genetics and feeding practices are working to produce lean meat.The Henry and Luthi farms are a striking contrast. Henry's operation markets 16,000 hogs a year from a complex of modern confinement buildings. That includes about 6,000 gilts sold for breeding. The farm produces the breeding stock under an agreement with PIC (former Pig Improvement Company), a hog-breeding company that is the largest in the world. The Luthi farm buys breeding stock from another company, Farmers Hybrid, and farrows its own feeder pigs from sows in dirt lots.

Some members of the group are considering starting their own farrowing co-op to produce high quality feeder pigs that will be consistently lean and will gain weight efficiently. They're also considering forming a co-op to have feed milled to their own specifications.

Another marketing approach that works for a few livestock producers is to bypass the system completely and sell directly to consumers. Don and Ruth Lowenstein of Cameron, Missouri, raise about 40 grass-fed calves a year that they sell as natural beef to consumers in the Kansas City area. In the spring of 1994 when Larry Anton saw the cattle market fall into the low $60 per hundredweight range, the Lowensteins had already sold their yearling calf crop for the coming fall to consumers for $1.60 a pound. That came out to between $80 and $105 a hundredweight if the calves had been sold on a liveweight basis.

Pat Steffen, the Nebraska beginning farmer featured in Chapter 1, was considering selling his hogs to a small, independent plant that wanted a source of "organic pork." Steffen already had the advantage of producing his hogs for about $35 per hundredweight, which was under the cost of many established farmers with more expensive facilities, including some of the most efficient hog producing corporations.

The changing economic structure of the hog industry and livestock markets may be the darkest cloud on the horizon for beginning farmers who raise cattle or hogs. Economist Rhodes doubts that co-ops, or at least the big regional co-ops like Farmland, will try to shelter independent hog farmers from structural changes in the industry, mainly because that role doesn't fit the philosophy of the big co-ops.

New, independent producers in this climate will have to know every cost they incur, know the latest about genetics and technology, and know when to reject the most costly, risky advice of the experts. And even with all that, there will be many obstacles for the small, independent farmer starting out with low-cost methods.

For example, hogs born in inexpensive outdoor huts and finished to market weight outdoors or in buildings with open-air ventilation need more fat under their skin to stay warm in winter. That backfat is vital insulation. But those hogs with a slightly bigger layer of backfat will be vulnerable to discounts from packers who buy most of their animals from large companies raising leaner hogs indoors.

Some young farmers are borrowing $100,000 or more to put up buildings where they will raise hogs on contract for large companies who retain ownership of the hogs. They may be buying into vertical integration. That can be a way to start out in farming. But attorneys familiar with standard contracts offered growers say that the potential return is smaller than for independent producers. The contracts give few rights to the producer and should be studied carefully, with professional advice.

Roy Henry admits to sharing some of Dave Williams' worries about the future of independent hog farmers. But he adds that larger hog-raising corporations will have higher labor costs because they'll have to pay for unemployment insurance, workmen's compensation and other costs independent producers may not incur.

And, even though the smaller producers may be no more environmentally conscientious than big hog companies, Henry expects the larger companies to bear the brunt of complaints about pollution. Smaller producers can spread the animal waste over more land than the large, concentrated hog operations. The smaller producer "won't spend time in litigation, which the larger producers will."

"We have more advantages than the larger people do if we work together," he says.

CHAPTER 4

A Stingy Uncle Sam: Of Billions Spent on Farming, Little Goes to Beginners

As head of FmHA for Iowa (now the Rural Economic and Community Development arm of USDA) Ellen Huntoon was one of several Midwestern state directors to use beginning farmer loans. In 1995 her state led the nation in loans to young farmers. Photo by Dan Looker.

On October 1, 1993, Tim and Kathy Pick and their son, Eric, found themselves pictured on the front page of *The Lincoln Star*, the morning newspaper in Nebraska's capital city.

What had they done to deserve their fifteen minutes of fame? They were the first farm family in the nation to get a "down payment" loan from Farmers Home Administration, the lending arm of the U.S. Department of Agriculture. The federal government loaned them $65,000 at 4 percent interest to pay 30 percent of the price of a 240-acre farm. They are buying it from Tim's uncle, who is financing 60 percent of the cost himself through a contract with the Picks. The young couple came up with 10 percent down on their own, after renting the diversified hog, cattle and crop farm for a few years.

"It's a good program for beginning farmers," the 28-year-old Tim Pick said later of the Farmers Home Administration down payment loans.

Although this was a good news story with some appeal to readers in one of the nation's most rural states, it was news precisely because such help for young farmers is unusual. Nebraska happened to be first to use the program because of a public-relations-style promotion campaign by Stan Foster, an energetic farmer and former Peace Corps volunteer appointed by the Clinton Administration to run Farmers Home Administration in the state.

Months before the government published the official rules for the program, Foster met with bankers, farm groups and his own staff to explain it. And he made sure that farmers knew about it through a radio, TV and newspaper blitz that generated 500 calls from interested young people across the state. The new program, signed into law by President George Bush in the fall of 1992, needed that kind of enthusiasm because it is more complicated than older government loan programs. Instead of the government making a loan directly to a young farm family for the entire mortgage on a farm, it loans only 30 percent. So the young farmer needs to find a bank or a landowner willing to finance 60 percent. And he or she needs a modest nest egg to put up 10 percent.

By the 1990s, buying a farm with only 10 percent down might have seemed like a throwback to the agricultural boom of the 1970s. But, in fact, it was a fairly conservative program. Other lending programs still on the lawbooks allowed Farmers Home to require nothing down at all to buy a farm and to stretch the payments out over 40 years. That credit seemed easier, but the borrower would be paying the government for half of his or her productive life and have little to show for it. Under the new program, as Nebraska Farmers Home director Foster points out, the farmer must pay back in 10 years the government's down payment loan for 30 percent of the farm's value. That's not an easy thing to do. But at the end of 10 years, with the government paid off, with the farmer's own 10 percent down, and with whatever else has been paid on the loan from the bank or landowner, the beginning farmer should own half of the farm. That's a 50-percent-equity position that would look good to the more conservative bankers of the 1990s. At that point, the young farmer could get commercial loans.

The program had another advantage. It allowed the federal government to help more young farm families with a smaller amount of money than if it made only direct loans to young farmers for all of a farm's price.

That kind of logic appealed to one of the program's chief sponsors in Congress, Rep. Tim Penny of southeast Minnesota, a conservative deficit hawk in the Democratic Party who favored steeper cuts in the federal budget than those offered by the administration of President Bill Clinton. (Penny retired from Congress in 1995.)

In spite of the apparent advantages of this new program, the Clinton Administration didn't seem to embrace it with enthusiasm. Penny and Rep. Jim Nussle, an Iowa Republican, wrote the new Agriculture Secretary, Mike Espy, twice in the administration's first year urging him to put the new law into effect faster. In September, a hearing by the House Environment, Credit and Rural Development Subcommittee extracted a promise from Espy's staff that the program would get more attention.

In the first three months of the federal government's fiscal year that began in October of 1993, Farmers Home Administration did almost nothing to promote the program. The hard work by its staff in Nebraska, Iowa, and a few other states was an exception.

An article in the March 1994 issue of *Successful Farming* showed exactly how rare Tim and Kathy Pick's experience really was, and how uninterested the federal government seemed to be in beginning farmers.

Congress had budgeted a modest $25.8 million for the program nationwide. In just over three months at the beginning of the federal fiscal year—from October 1, 1993, through January 12, 1994—Farmers Home Administration had spent only about 2 percent of that amount and had made only 14 down payment loans. Only seven states had used the program—Kansas, Oklahoma, Oregon, Pennsylvania and South Dakota in addition to Iowa and Nebraska.

By April, after the winter land sales season, the agency had increased the number of down payment loans it had made to 107, but it appeared that it would fall short of the goal set for it by Congress. And 20 states still had not used the program. Congress had also set an April deadline for the USDA to establish an advisory committee for beginning farmers and ranchers. "Nothing has been done at all," Nancy Thompson, an attorney for the Center for Rural Affairs in Walthill, Nebraska, said a few months after the deadline.

In recent years, the nonprofit advocacy group founded in 1973 has made beginning farmer programs the cornerstone of its work on behalf of family farmers. The Center lobbied hard for the down payment loan program that had been passed by Congress. But in the program's first year, mere mention of it seemed to draw a sigh of resignation from Thompson.

"It's kind of ironic that you had to create beginning farmer programs within Farmers Home Administration," Thompson argued. After all, the agency grew out of a New Deal program established during the Depression of the 1930s to help sharecroppers get started in farming. Known as the USDA's "lender of last resort," it was supposed to help farmers who could not get credit elsewhere, a common situation for young farmers of modest means. And the agency has, in fact, helped many farmers get started over the years. But, by the 1990s, it was laboring under years of bad publicity, conflicting goals ordered by Congress, and in 1994, Agriculture Secretary Mike Espy's goal of reorganizing the USDA.

"Everybody is so preoccupied with rearranging the chairs that they're ignoring implementation of programs," said Thompson.

Thompson also wondered if either Clinton or Espy had much interest in beginning farmers, even though both professed a lot of interest in rural economic development. Espy also said in a January 1993 interview with *Successful Farming* that "we have to make farm life appealing to the young ... showing them that the government will be helpful to them on alternative crops ... making sure that they get a friendly hand at the commercial bank if they can't get a direct loan [from Farmers Home Administration]."

But Espy's interest in young farmers seemed less intense than his desire to promote nonfarm rural development. Under his plan to reorganize the USDA, Espy was splitting Farmers Home in two. He was putting loans to farmers in a new Farm Service Agency that rolls most of USDA's farm programs—price supports, crop insurance and loans—under one roof. Farmers Home Administration's other loan programs for housing and nonfarm programs was to be shifted to a new Rural Development Administration.

"I don't think either Clinton or Espy understand the relationship of developing rural communities and helping beginning farmers," Thompson said. "They're ignoring one of the key components."

Another problem for Farmers Home Administration is that the agency had become a public symbol of the corruption and misuse of programs that sometimes plagues the USDA.

"Every time there is a new scandal, there is less and less interest in Farmers Home" in Congress, Thompson said.

One of the latest was a report in *The Washington Post* in January 1994 that described an apparently wealthy southern California dentist who owes Farmers Home Administration $3.5 million and has been delinquent on his payments 13 years. The story goes on to describe other California borrowers who would hardly fit the common stereotype of a family farmer and it points out that the agency wrote off $11.5 billion in bad loans from fiscal year 1988 to 1992. It portrays the agency as so tolerant of delinquent borrowers that its loans are really grants.

Buried deep in the story is the fact that the agency's more generous programs have been discontinued and the article never really makes it clear that none of the uncollected big loans could be made today. It describes a Congress that became even more lenient about collecting loans in a 1987 law, but it ignores the historical context. In that year, farmland values had fallen to less than half of their 1981 value and commercial lenders weren't eager to foreclose on their borrowers, either. They, too, were "restructuring" loans of their farm borrowers. Often, allowing a farm to continue operating with some debt forgiven or extended over longer terms was preferable to taking over the farm and putting its devalued land on the market.

Thompson was frustrated by the fact that the story ignored that delinquency rates are low for the direct Farmers Home loans targeted to family farmers. And those loans are limited to no more than $200,000.

Nevertheless, the investigation by the *Post* had turned up yet another example of abuse of Farmers Home programs that long plagued the agency. And it wasn't always just the borrowers who misused the government program. In the 1980s, Norm Brewer wrote a series of reports for *The Des Moines Register* about allegations by farmers who accused the Eastern Iowa Production Credit Association (a member of the farmer-owned private lender, the Farm Credit System) of fraudulent use of guarantees of the PCA's loans to farmers. In recent years, Congress has de-emphasized loans made directly by the agency, instead asking Farmers Home to guarantee private bank loans. If the loans go sour, the government reimburses the lender for nearly all of the losses.

Some Iowa farmers alleged that the PCA made FmHA guarantees a requirement of getting their own troubled loans renewed in the 1980s. In some cases, they alleged, the PCA used appraisals that overvalued their land. Shortly after getting the guarantees, the PCA foreclosed on the farmers and collected the guarantees. Although a grand jury investigation of the allegations never resulted in criminal charges against the PCA loan officers, the Justice Department in 1992 reached an out-of-court agreement with the now defunct PCA's successor and its parent bank, The Farm Credit Bank of Omaha, to repay the government $4.2 million.

During the new conservative mood of the 1990s, when the public and Congress were looking for ways to trim unneeded federal programs from the federal budget, Farmers Home Administration might have seemed a tempting target. Thompson and her colleagues at the Center for Rural Affairs were strongly opposed to doing away with the agency because it was still the best chance to offer at least some meaningful help from the government to farmers who most needed it, especially beginning farmers.

Yet, the agency's defenders have for years been one of its strongest critics. In his book, *Family Farming: A New Economic Vision,* the center's co-founder, Marty Strange, clearly describes Farmers Home Administration's evolution to a bloated, inefficient mechanism for meeting the changing whims of Congress on farm policy.

Farmers Home Administration can trace its roots to the administration of Franklin D. Roosevelt, who created the Rural Resettlement Administration to provide acreages for farming to dispossessed tenant farmers during the Depression. It was renamed the Farm Security Administration and made operating loans to tenants to strengthen their bargaining position with landlords. In 1937 Congress authorized the Farm Security Administration to provide 40-year loans for tenants, farm laborers and sharecroppers to buy land. It was an activist role clearly aimed at the most disadvantaged farmers. But, as Strange points out in *Family Farming*, it began to change in the 1940s; "With farm prices rising, the zest for agricultural reform had given way to milder ambitions." The agency began making loans to existing farms to expand and modernize. The old Farm Security Administration was changed to Farmers Home Administration in 1947. Still, throughout the 1950s, the agency's role was targeted at modest farms. "It was to provide *interest-subsidized loans in limited amount to family-sized farmers who could not get credit elsewhere but who, with management assistance, could be expected to graduate to commercial credit,*" Strange writes.

However, Congress also used the agency to respond to crises and political pressure from farmers. During the Depression Congress created the Disaster Loan Program to help farmers recover from natural disasters. In 1953, it expanded the program to cover economic disasters, such as low commodity prices. These loans later became Economic Disaster loans, which had no loan size limit and looser eligibility requirements. In the 1970s, "big, rapidly expanding farms found the FmHA door open, or at least unlocked," Strange writes.

The expansion of the 1970s led to higher production, higher machinery and fuel prices and other forces that began to diminish profits—especially for those with high debts—as early as 1977, when the American Agriculture Movement threatened a national farm strike. Congress responded in 1978 with yet another credit bonanza, the Economic Emergency loan program. In his book, Strange writes:

> In startling contrast to the agency's traditional purposes, loans were no longer limited to family-sized farms. To qualify, borrowers did not even have to rely on farm income for their livelihood. ... Under the EE program, borrowers could get up to $400,000—twice the amount available for the agency's regular real-estate loans, and four times its regular operating loans.

Strange, citing statistics gathered by former Center for Rural Affairs attorney Gene Severens, shows that by 1979 most of the agency's outstanding loans were made to farmers bigger than its traditional clientele. During the farm debt crisis of the 1980s, the Reagan Administration virtually abandoned any pretext of helping the most disadvantaged farmers. Congress eliminated the lax economic emergency loan program in 1984, roughly ten years before *The Washington Post* was exposing it as a still-wasteful boondoggle. But the Reagan Administration also shifted many larger farms into the regular programs for disadvantaged farmers.

Strange has been critical of the government's broad-brush use of Farmers Home Loans during the 1980s and he admits that his point of view hasn't always been popular with farmers.

At The First National Conference for Beginning Farmers and Ranchers in Omaha in March 1994, Strange said, "There are plenty of reasons not to help beginning farmers. During hard times, we are told we have to save the farmers that are already out there, not help new ones get started. In the good years we are told, 'Why worry? No one needs help.' "

"The moral flaw in the family farming system is that too often we have been so concerned with the vested interests of people who own farmland or might inherit it that we have ignored the need to bring new blood into farming," he said.

Looking at the government's farm policy in general, and other farm programs besides Farmers Home Administration loans, it's clear that farm policy has responded to those who have the most clout in Washington, organizations that represent different commodities produced by farmers and general farm organizations that represent established farmers. It's doubtful that any of those groups consciously want to exclude young farmers with modest resources. But it's obvious that those young farmers or would-be farmers are often overlooked.

Commodity programs that help support the cost of growing crops such as corn, barley, wheat, cotton and rice, would seem to benefit all farmers equally. But numerous studies have shown that bigger farms tend to get more of the payments from these program. Congress has placed limits on these payments—currently at $50,000—but there are still ways around the limits, through partnerships, for example.

Commodity programs also have a tendency to increase the cost of land and make entry into farming more difficult. In the Midwest, a key question farmers ask about a farm that is put on the market is the size of its "base"—the acreage with a history of growing crops eligible for income support payments. The bigger the base on a farm, the higher its price.

Until recently, farm programs offered the income security that allowed existing farms to expand by borrowing to buy more land. In a way, federal farm programs have worked in tandem with the more powerful "treadmill of technology" to speed farm consolidation. To Strange and other thoughtful advocates of dispersed land ownership by modest, middle-class farmers, federal farm programs are seen as highly flawed. But to the average farmer struggling to make a living raising corn or wheat on a modest-sized farm, the complicated commodity programs are a necessary evil that brings in a little more income. It can be compared to Congress voting to lower federal income taxes. If wage earners get a larger than expected refund from their tax withholding, no one complains. But, in the long run, if that tax break helps increase the federal deficit, we all may pay for it many times over as higher interest rates ripple through the economy.

By the mid 1990s, many farmers active in commodity organizations knew that traditional farm programs were nearing their demise. In Iowa, a committee representing nearly every farm group in the state studied their options and came up with a proposal for a new farm program that they called "Revenue Assurance." It would do away with commodity programs, disaster payments that Congress faithfully approves after every drought, hurricane or flood, and federal subsidies for crop insurance. Instead, it would combine those federal funds into a less-costly guarantee that a farmer's income wouldn't fall below a certain level, say 70 percent, of the average of cash receipts for the previous five years. That meant, in effect, that farmers would only be paid in years of natural disasters, not every year as they are now. The plan recognized that the federal government would be willing to spend less on agriculture than in the past.

Doran Zumbach, an Iowa corn farmer who helped organize the group that came up with what was known as "The Iowa Plan," said that his group had talked a lot about what might be done to help beginning farmers with the new system. One idea was that the government would guarantee a higher level of income, say 80 percent, for the first ten years of a farmer's career. Significantly perhaps, the group didn't put that idea in the proposal as it sought support from farm groups in other states.

The Iowa Plan and *The Washington Post* article on Farmers Home Administration may be signs of a new conservative mood on farm policy that could throw the baby out with the bathwater as Congress struggles to write a new farm bill in 1995. Privately, many in Congress see no need to help young farmers get started when they think the nation still has too many farmers.

Not everyone was cynical about beginning farmer loans, however. Dedicated public servants like Ellen Huntoon, who became one of FmHA's first state directors in Iowa, were well aware of the agency's tarnished image after more than a decade of heavy loan losses.

Under Huntoon's leadership, Iowa surpassed Nebraska as the biggest user of beginning farmer loans. During the agency's 1995 fiscal year, when it made more than 800 direct loans to Iowa farmers and guaranteed more than 1,100 bank loans, more than 200 borrowers were classified as beginning farmers.

"It seems to me very important for Farmers Home to get back to its original mission. It was created to help people start farming," she says. So Huntoon has directed her staff to seek out potential beginning farmer borrowers. The fact that the agency already has a good working relationship with banks that use its guarantees has helped her staff cooperate with private lenders to find those young farmers, she adds.

Under the reorganization of the U.S. Department of Agriculture by the Clinton Administration, the agency was to be split in two during 1995. Farm lending by FmHA would be shifted into a new consolidated Farm Service Agency, whose main role would be to administer the price support programs formerly run by the Agricultural Stabilization and Conservation Service. The rest of the agency would become a new Rural Economic and Community Development arm of USDA.

Not only was the agency being split in two, but some members of the new Republican Congress had begun to question its usefulness. One of the questions asked by the new Senate Agriculture Committee Chairman, Richard Lugar (R-Ind.), shortly after he took over leadership of the committee was, "Is the Farmers Home Administration needed to encourage replacement of retiring farmers?"

That view that the nation has more than enough farmers to replace retiring farmers could lead to a tragic mistake—and a marked contrast to the help that those who are now retiring received from the government. After World War II the federal government encouraged returning veterans to start farming again. They were even paid to attend special classes in agriculture in the early 1950s. Today, an entire generation of farmers who started out in the 1950s and 1960s is nearing retirement. As we'll see in the next chapter, the potential loss from failing to plan for the next generation of farmers will reach from the main streets of small towns to urban consumers.

Washington's shortsightedness is nowhere more obvious than on the farm of Pat and Julie Steffen, described in Chapter 1. In the same year when rich borrowers of Farmers Home Administration loans were fighting the government's efforts to collect, Pat told young farmers at The First National Conference for Beginning Farmers and Ranchers that his business had done so well that in three years he had "graduated" from the subsidized interest rate the agency charges "limited resource" farmers.

"We're going to have to pay more interest but it says something, the fact that in three years we've gotten ourselves graduated," he said. "I'm fairly proud of that."

Pat has other things he can be proud of, too. By switching his farm's cropland to grass he is conserving the farm's precious topsoil. And the cattle and hogs he raises are of such high quality that they could be sold as "organic"—raised without hormones or antibiotic feed additives.

Yet, the paint on their large two-story farmhouse is peeling and the home is sparsely furnished. It's a poignant contrast to the "farmer" described by *The Washington Post* as a Farmers Home Administration borrower. The wealthy California dentist had an $817,000 ocean-front house with floor-to-ceiling windows.

Unlike some of California's wealthiest farmers, Pat Steffen is a purist who has no desire to milk federal programs to supplement his income. The nation has thousands of potential beginning farmers like the Steffens who don't want a permanent handout from the government, who are willing to forego comfort and consumer goods to get started raising quality food. If they take on debts, it's with an almost reverent sense of obligation. All that some of these farmers need is a modest loan from a sometimes confused but still resourceful Uncle Sam.

CHAPTER 5

Why Support Beginning Farmers? The Public and Small Towns Have a Stake

Former presidential adviser and congressman Cooper Evans of Grundy Center, Iowa, believes the government could do more to help beginning farmers at very little expense to the taxpayers. Photo by Michael Malone, courtesy of *Successful Farming*.

Jeff Heil is just about as careful with herbicides on his central Iowa farm as anyone. He uses "ridge tillage" to grow 650 acres of corn and soybeans on his 800-acre farm. The crops are planted into a raised seedbed, or ridge, in the spring. Planting lowers the ridge slightly. Later, as he cultivates to control weeds, he builds it back up. This system allows Heil to spray chemicals in a "band" over the seedbed. He doesn't spray the entire field. He even built a protective "hood" over his sprayer nozzles to keep wind from blowing herbicides away from the seedbed.

Heil, a good businessman who wrote his own accounting system for his computer, doesn't stop there in his effort to hold down chemical costs. The 37-year-old farmer mixes herbicides with soybean oil and an emulsifier. The "encapsulated" herbicide sticks better to the leaves of weeds, making it more effective. So Heil can use only half the manufacturer's recommended dosage for post-emergence herbicides sprayed after the crop is

43

growing. He can cut his pre-emergence herbicide dosage by about 30 percent. To make sure that tiny amounts of chemicals control weeds, he sometimes drives his tractor and sprayer in the evening, so the herbicides will soak in before sunlight can start breaking them down.

I talked to Heil about his system in 1990 for a series of articles that I wrote for *The Des Moines Register* called "A Lost Generation: Iowa's Vanishing Young Farmers." At the time, few of his neighbors were so conscientious. The older farmers had their land paid for and didn't have to watch expenses quite as closely, he said. So they just followed the company recommendations. It's simpler. And if the product doesn't control weeds and the farmer seeks compensation from the chemical company, he's on firmer ground. Older farmers still using full rates are following the environmental rules, he adds, "but they're going right up to the limit."

Today, Heil finds more of his neighbors using some of his new methods, especially the soybean oil encapsulation. "Quite a few farmers in the area have picked up on this. The majority are the younger ones," he says, although a few older farmers are trying it, too.

Heil believes that the younger farmers, partly because of their need to cut costs, are much more receptive to new ideas about ways to use fewer farm chemicals. And they're sensitive to environmental issues, too.

Jeff Heil is a good example of the stake that the public has in seeing more young farmers on the land. There is some scholarly evidence that young farmers are more inclined to try farming methods that fall under the broad category of "sustainable agriculture." There are a lot of definitions for sustainable agriculture, ranging from farming that uses almost no products purchased off the farm, including organic farming, to methods like Heil's that make judicious use of newer, perhaps safer pesticides. In this sense, the public, which has a strong interest in preserving the environment, may gain from policies that encourage young people to start farming. Young farmers could play a role in making certain that rivers and groundwater continue to provide high quality drinking water and that food remains wholesome and healthy.

Not only are young people more receptive to new ideas in general, but beginning farmers are at a stage in their lives when they're making decisions about the kind of machinery they will buy and the methods they will use. For older, established farmers, drastic changes in farming may require a costly investment that they are unwilling to make when they're starting to plan for retirement.

Younger farmers also need to maximize their income from sales and may be more inclined to bypass the traditional marketing and processing system. Every year Heil, for example, plants five to ten acres of sweet corn that his four children help him harvest. Sometimes they sell the crop at a farmers market in nearby Marshalltown. Heil makes a profit that is "unreal compared to commercial corn"—$500 to $700 an acre compared to $50 to $70 for field corn. And the consumers of Marshalltown get to buy fresher, sweeter corn that hasn't been shipped halfway across the country. Whether young farmers sell free-range chickens in the East, lean "freezer beef" in the Midwest, or fruit grown with reduced pesticide use in California, they are an important link between consumers and agriculture.

Younger farmers also may have less land (depending on how much help they got from their parents) so the ones with smaller operations may have more time to use sustainable methods. Jeff Heil's system of ridge tillage and sometimes applying herbicides at dusk wouldn't work on a huge acreage. "You have to be extremely timely when you start cutting back on herbicides," he explains.

Another type of new technology was sweeping the Corn Belt in the 1990s. It's called *no-till*. It relies on herbicides for weed control and uses heavy duty planters and drills to put seeds directly under residue left on the ground from the previous year's crop. Like ridge tillage, no-till is an improvement over older tillage methods because it helps prevent soil erosion. But it doesn't offer quite the same opportunity to cut back on chemical use. It's being adopted by some of the very biggest, established farms, Heil says. "The reason it's going over so well is that these larger operators can farm more ground, yet."

The general public probably isn't even aware of the importance of a new generation of farmers. But the loss of farms and its effect on rural communities is a frequent conversation topic among farmers and people in small towns.

The importance of beginning farmers to the fabric of rural America is obvious to Cooper Evans, a former Republican congressman from Grundy Center, Iowa, who served in the Bush Administration as Special Assistant to the President for Agriculture.

"I think there is a broad recognition in the country that poverty is not confined to the cities," Evans says. Because of poverty in rural America, there is a need for rural economic development.

"A very important part of that needs to be built on the agricultural resources of rural America," he says. Land, and to some extent, the buildings of farmsteads, remains the most important asset in rural communities. And, while it might sound improbable to those who live outside rural areas, there is a risk that much of that land won't be used by the largest commercial farms.

"If there isn't 10,000 acres there, even commercial agriculture isn't going to be interested in it," he says. "A beginning farmer program can recognize the existence of those assets." On his own farmland near Grundy Center, Evans has practiced what he preaches. He has rented land to several young farmers to help them get started. And on bottomland along the Blackhawk Creek, east of town, he has planted his own alternative crop, black walnut trees.

There's more to rural development than promoting high-technology industries and seeking factories to relocate in rural areas, Evans says. A beginning farmer program ought to be part of the federal government's rural development efforts—"and I think that it is overlooked."

That viewpoint draws strong agreement from Marty Strange of the Center for Rural Affairs. "The beginning farmer is a new business and that's how it ought to be viewed by the community," he says. "Small communities depend on people, not on commodities." Beginning farmers will buy more than seed and fertilizer. "They buy food and clothing and consumer items that only get sold to people," he says.

But as the farm population has decreased in America's rural, farm-dependent counties, many small towns without other nonagricultural industries have boarded-up main street shops and little retail trade. Often the biggest business in a small town is the grain elevator. "Small towns have been transformed from a rural community to a wholesale and warehouse district," Strange says.

The relationship of small towns to the kind of agriculture that surrounds them—the size of farms and how they are owned or controlled, or the "structure" of agriculture—has been fertile field of study for agricultural economists and rural sociologists. And there is a lot of scholarly work that seems to support the views of Evans and Strange.

Probably the most famous is a 1944 study of two California towns in the San Joaquin Valley, Arvin and Dinuba, by a USDA economist, Walter Goldschmidt. He knew that agricultural production was becoming concentrated in fewer hands and he wanted to test the effects of this change in structure by comparing two communities that were similar in most ways, except that one had fewer farms nearby. Dinuba had 722 farm operators with an average farm size of 57 acres. Arvin had 133 farm operators with an average size of 497 acres. The smaller farms near Dinuba were more typical of what we might consider to be family farms today. More than three-fourths of them were run by full owners. Only about a third of the farms surrounding Arvin were run by owners. Goldschmidt found more farm laborers in Arvin and more farmers and white-collar workers in Dinuba.

The business sections of two California towns, Arvin (*above*) and Dinuba. Photos by Dan Looker.

He concluded that several measures of what we today call "the quality of life" were higher in Dinuba, the family farm community. It had more schools, parks, churches, civic groups and paved streets. It had more than twice as many retail stores and a bigger trade volume.

The study was controversial and was suppressed. Goldschmidt was fired from the USDA. Years later, in 1968, he told a committee of the U.S. Senate that smaller-scale farms produce a middle class that has a "strong economic and social interest in their community. Differences in wealth among them are not great, and the people generally associate in those organizations which serve the community."

Goldschmidt's study of the two California communities has been updated in recent years by others and it also spawned a whole field of scholarly work. Much of Goldschmidt's conclusions have held up under scrutiny, although some scholars have questioned whether Arvin and Dinuba were really comparable towns. And in recent years some have criticized the methods and interpretation of his work. Ohio State University economist Luther Tweeten, for example, has pointed out that the farms around Dinuba were middle-class farms in 1944, not small farms by the standards of the day. And, he adds, the standard of living was not so high in many towns in the South that served truly small farms. An excellent review of this scholarly work was published in 1990 by the State University of New York Press. It is *Locality and Inequality: Farm and Industry Structure and Socioeconomic Conditions* by Linda M. Lobao.

The relationship of sustainable agriculture to farm size and a farmer's age is a different type of scholarly pursuit that has begun only recently. It seems to support Jeff Heil's common-sense observations in his own rural community. One study published in 1992 by Gordon Bultena and his colleagues at the Iowa State University Department of Sociology showed a relationship between age and adopting sustainable farming practices. In a study of more than 1,000 Iowa farmers, Bultena found that the average age of farmers "in transition" to sustainable farming—40—was younger than the average of 48 for the entire group. But those already practicing sustainable agriculture and those using conventional methods had about the same average age as the whole group.

Bultena adds that 40 percent of these farmers began adopting more sustainable practices before the age of 30, and 76 percent before the age of 40. Only 8 percent began adopting after age 50.

The same study, *Transition to a More Sustainable Agriculture in Iowa* (Sociology Report 166), also found that conventional farms were the largest, averaging 673 acres, while farms in transition averaged 578 acres and the sustainable farms were the smallest, averaging 391 acres.

A 1993 study by Rebecca S. Roberts of the University of Iowa Department of Geography showed a clear difference between the willingness of older and younger farmers to adopt ridge tillage and herbicide banding. In-depth interviews with members of 40 Iowa farming operations showed Roberts that there was "resistance of older farmers in their 50s and 60s to innovate with low-input practices such as sidedressing, herbicide banding, or ridge till. Older participants made less frequent use of these practices and were more likely to perceive low-input practices as too risky, infeasible on their farm, or outdated."

Ultimately, though, the worth of beginning farmers may be unprovable. Neither economics nor sociology are exact sciences that can test theories in a laboratory. Whether rural America and society in general decide to encourage more beginning farmers will be a subjective, political value judgement.

Nor will beginning farmers be a panacea that solves all of the unpleasant side effects of modern agriculture. They won't completely stop the treadmill of technology that decreases the overall need for farmers. And the group marketing that some young farmers use to survive as agriculture evolves toward more vertical integration and economic concentration may bypass small-town livestock auctions. That might speed the demise of some rural communities, or at least some small-town businesses.

The least likely person to portray any change in agriculture as a panacea is Marty Strange. A witty, articulate advocate of family farmers for more than two decades, Strange is regarded even by critics as a brilliant advocate of a Jeffersonian vision of equal opportunity in farming. He and his co-workers at the Center for Rural Affairs have lobbied Congress to eliminate some federal income tax breaks that encouraged excess spending in farming by wealthy, absentee investors. They have encouraged scholarly research on sustainable agriculture. They offered well-documented criticism of university research that ignores the needs of smaller producers. In Nebraska, they helped win passage of a state constitutional amendment that bars nonfamily corporations from common forms of agricultural production. They also started Land Link Realty. It's a program that puts retiring farmers in touch with unrelated young farmers interested in taking over independent, sustainable farms.

Yet, the Center's influence on farm policy hasn't slowed the inexorable trend toward fewer farms. It seems almost powerless to slow the trend toward more market power by the large agribusinesses that supply commercial farms and buy their products.

Marty Strange still dreams of seeing more farms in the United States but has few illusions that his dream is likely to be realized soon. "We're a long way from hoping to do that, but at least we can slow the decline—and that's what the beginning farmer movement is all about," he says.

Neither Strange nor Evans, two of the most prominent advocates of beginning farmers, sees quick recognition in Washington of the need to help beginning farmers.

"I'm not encouraged by the attitudes of the Clinton Administration, not at all," says Strange, who finds little interest in beginning farmer issues there. At The First National Conference for Beginning Farmers and Ranchers, Evans said, "There are too few farmers to have much political clout" in seeking ever scarcer federal dollars. To make matters worse, "The public image of those of us who farm [is] not what it used to be."

The solutions the two men offer differ. Strange favors a more activist approach by government, including the judicious use of credit by Farmers Home Administration. Evans has a long list of things the government could do that focus more on farm programs and tax policies.

"There really are many innovative ways government could facilitate the process of succession and entry into farming," Evans said in his speech at the Omaha conference for beginning farmers and ranchers. "There could be a partial tax exemption on income from renting land to a beginning farmer. There could be a partial tax exemption on interest received from selling land on contract to a beginning farmer. The capital gains tax on land sold to a beginning farmer could be reduced. Property taxes could be lowered on land owned or operated by beginning farmers. We do this for businesses all the time. The commodity loan rate could be increased temporarily for beginning farmers, then gradually brought back to the standard rate."

Despite the gloomy political prospects for farmers in Washington in the 1990s, Evans did see the possibility of "a rather broad coalition of what I'll call strange bedfellows committed to addressing the problem."

"Concern about declining farm numbers and the number of farmers and the difficulty of entering farming is beginning to emerge in unexpected places, places that were little interested or gave little sympathy in the past," he said. "Giant operations, such as Murphy [Family] Farms or Premium Standard Farms [among the nation's largest hog companies], ... or even very large family operations don't really have much need for the Farm Bureau, for example. Or the National Pork Producers. Or the Corn Growers. Or the Soybean Association. And these giant farmers certainly don't provide as many members to such organizations. And they don't need farmer-owned cooperatives or local agribusiness. They may not even need Cargill or Continental [Grain Company]."

Strange sees the real battle for beginning farmers being waged in that potential rural coalition. "These battles are not won and lost in Congress. The pressure has to come from out here. Congress only reflects public attitudes," he says.

"If established farmers see beginning farmers as an economic threat, they will become the biggest political enemies of beginning farmers," he adds. Those most threatened would be the ones who want to buy a lot more land to expand their own farms. But most farmers probably have mixed feelings. "They would like to have more neighbors and would like to have more land."

Besides all of the benefits that beginning farmers offer rural communities, the environment and urban consumers, they also play a key role in making the United States competitive in the global economy, adds Strange. "The competitiveness which characterizes the American farm economy is the product of the opportunities created by the system. If all of the resources of a sector of the economy are locked into a few hands, you almost always get economic lethargy."

Many Americans can remember the kind of "economic lethargy" that produced gas-guzzling, breakdown-prone cars when three big automobile companies in Detroit dominated the industry. Without the competition that came from Japanese imports, the U.S. auto industry "would be in the dark ages," Strange says.

Agriculture remains one of the most competitive sectors of the U.S. economy and is a long ways from being completely dominated by three companies. Whatever the political future holds for beginning farmers, each one will ultimately succeed or fail individually. The forces of politics, government and the economy will affect those farmers, but just as significant will be each farmer's ability, training and desire to farm.

The next chapters show that the federal government's scant attention to beginning farmers isn't reflected in the countryside. State and private programs are springing from the grass roots to encourage young farmers. And young people are succeeding in getting started in farming in spite of all of the barriers and ominous changes in contemporary agriculture. Each of these new farm families has many reasons to be proud of their role in agriculture and their contribution to the nation.

What's There to Help Young Farmers?

CHAPTER 6

Matchmaker, Matchmaker:
New Links Between Older and Younger Farmers

John Baker, who runs Iowa's successful Farm-On matching program, goes over some estate planning ideas at a conference for beginning farmers sponsored by *Successful Farming* magazine. Photo by Ed Lallo, courtesy of *Successful Farming*.

In 1994, the Iowa legislature created a new "Beginning Farmer Center" for Iowa State University and the state's Cooperative Extension Service. In spite of a tight state budget, the legislature appropriated $100,000 to start the center. It coordinates efforts to help beginning farmers and makes annual reports with "recommendations for methods by which more individuals may be encouraged to enter agriculture."

The new law got little fanfare but may turn out to be a Magna Carta for a generation. "It makes it the state of Iowa's public policy to encourage young farmers," says John Baker, a Des Moines attorney who runs Farm-On. Farm-On is a joint effort by Iowa State University's Cooperative Extension Service and the Iowa Department of Agriculture and Land Stewardship. It keeps a computer database with names of older farmers looking for young people to work into their businesses and young people looking for a place to farm. It is just one of nearly 20 state or regional programs that have sprung up since the first matching service, Land Link, was started in 1990 by the Center for Rural Affairs in Nebraska. Some are private. Many are part of state departments of agriculture. This grassroots movement had the support of farm organizations, especially state Farm Bureau federations. In Iowa, Farm Bureau helped Baker recruit older farmers to attend Farm-On's workshops on estate planning. In Michigan and in Illinois, Farm Bureaus were looking into running matching services in those states.

Iowa's new Beginning Farmer Center and the support it got from the legislature strengthens that state's matching service. And it may symbolize a new attitude in the agricultural establishment of state universities, farm organizations and agribusinesses. After decades of policies that, whether spoken or not, seemed based on the assumption that the nation has too many farmers, some state governments seemed to be saying "enough" to downsizing in rural areas.

The concept of a matching service was ingenious. It requires very little money compared to most state or federal programs for farmers. Until the matching programs came along, the only government services for young farmers, besides educational programs, were loans from the federal Farmers Home Administration and, in some states, "aggie bond" programs. The tax exempt aggie bonds are used to lower interest rates on bank loans or contract land sales to young farmers. The new matching programs run at a fraction of the cost of these older programs. And, while they don't necessarily eliminate the need for loans, they have the potential to make entry into farming less costly for young people. They also can increase retirement income for older farmers.

Matching programs work something like a computerized dating service. Older and younger farmers fill out questionnaires. The older farmers describe the enterprises on their farms. The young farmers describe their experience, training and interests. A matching service might give out lists of names of farmers who raise cattle, for example, to younger farmers interested in raising cattle. Then, in the Nebraska program, the younger farmers contact the older farmers, who, in essence, conduct a job interview. Often the younger farmers start out by working as a hired hand, with at least a verbal understanding that they later may be able to buy a farmer's livestock, for example, on shares (keeping part of the offspring as payment for caring for the livestock). Less commonly, young farmers with experience gained by renting farms will buy the older farmer's business if the young farmer has some savings to make that possible. The farm might be sold on contract, with the older farmer holding the mortgage and taking annual payments from the younger farmer.

To Baker, who earned a master's degree in business and worked for a labor union before becoming a lawyer later in life, it seems odd that more farmers don't sell a working business to someone else. More commonly, if a farm couple has no children who want to farm, at retirement they'll auction off their machinery and liquidate any livestock they own. Then they sell their land, or rent it to another farmer. Either way, the farm is sold off in pieces, almost as if every retirement was a bankruptcy liquidation. In fact, many sales held after the debt crisis of the 1980s were liquidating healthy businesses. For most of the 20th century, the farm sale was a rural institution, drawing neighbors from miles around to buy or just to follow the prices paid for used machinery.

"I don't think that most businesses are sold that way. Most businesses are sold intact," Baker says. If a trucking company is sold, for example, "they don't sell the trailers to somebody and the engines to somebody else. The simple fact is that the business is more valuable intact than it is apart."

It might seem that an intact, operating farm would be too expensive for a young farmer to buy since it's more valuable. But that's not the case, Baker says. If a farmer sells everything, including his land, capital gains taxes will take so much of the proceeds from the sale that the income from putting what's left into the bank is often much less than the retiring farmer expected. Instead of doing that, an older farmer could lease his livestock buildings to a young farmer for a modest fee, for example, and either sell the young farmer the land through a contract or lease the land. The retirement income would be much higher than from most other alternatives, Baker argues.

One reason more farmers don't do this, he says, is "that as you get older, you tend to be risk averse. And, in an effort to reduce risk, they're paying an awful price." Working an unrelated person into a business might seem risky. But even those farmers who have decided to sell off the livestock and machinery and rent their land to an established farmer can't escape all risks. Baker recalls that in late 1993, when flooding and wet weather cut Iowa's average corn yield to 80 bushels an acre (even less than after the drought of 1988) many retired farmers didn't get the second half of their annual rental payments from established producers.

Those older farmers who work with Farm-On to bring a younger farmer into their business usually have two reasons, Baker says. "One is what I would call community motivation. That's the motivation to preserve their community," he says. "The other is that as they have looked at the alternatives to retiring; this is the best alternative economically."

Any kind of business relationship between older and beginning farmers has to work financially. But above all, the key to making any given match succeed is the personalities of the farmers and their spouses—"communication and trust," as Baker describes it.

Allen Prosch of the Center for Rural Affairs' Land Link program agrees. When he works with older and younger farm families, "I've learned that it is the personal relationship facilitating that is the critical part of these programs," says Prosch, who has the most experience with putting farmers with similar interests into business relationships.

Prosch spent more than the usual amount of time working with an eastern Nebraska farm where one young couple tried farming and eventually decided that it wasn't for them. Yet, another, slightly older young couple now live and work on the same farm and feel as if they've found a great opportunity. The farm makes a good case study for how matches between unrelated farmers really work.

The 800-acre farm of Del and Merle Akerlund is in Valley, Nebraska. Del, who is now 72, is one of the best-known organic farmers in the nation. His farm has been studied by the U.S. Department of Agriculture and by scientists for several scholarly research reports. Del started farming at ten, when he helped his father by driving a mule and a one-row cultivator through the corn fields. In 1948, when his father died, he and his brother began running the farm. Del farmed like everyone else in his neighborhood until 1967, when he decided to give up using pesticides because of his concern about the environment and health. He had some advantages over others who have tried organic farming. The town of Valley isn't far from Omaha, where he was able to get an inexpensive supply of natural fertilizer—manure from meatpacking plants. Over the years, that supply of manure has built up the organic matter in the soil of his farm, which lies in an already productive valley of the Elkhorn River. Del's good management and reputation for quality has allowed him to demand exceptional prices for his crops. He sold his 1993 soybeans for $17 per bushel for export to Japan and Europe. It's a special variety, Vinton, that's used to make tofu. The best that conventional farmers got for soybeans in 1993 was around $7 per bushel.

Most of the farms listed in the Land Link program aren't organic, but Del's reason for trying the program is similar to the motives of many other older farmers in state linking programs. He wanted to see his life's work carried on. "It would be so easy to turn it over to a conventional farmer," he says. "But I can't see it ruined. I would hate to see 25 years of organic farming go down the tubes. This is why I've been looking so hard for someone who really loves the land."

The Akerlunds decided to list their farm with Land Link to find someone who shared their philosophy. They have two daughters, but they aren't interested in farming.

At about the same time, Allen Prosch was holding classes for young people interested in farming. The classes covered setting goals and making realistic plans for running a business. In those classes Allen got to know a young couple from Ohio, Marty and Mindy Hitchcock, who had moved to northeast Nebraska, where Marty worked at the parts counter of a John Deere dealership. Marty had a strong interest in organic farming, and Prosch thought the Hitchcocks would make an ideal couple to work on the Akerlund farm.

Marty Hitchcock grew up on a small farm 18 miles from Cleveland, Ohio. His family owned only 30 acres, but his father finished hogs and usually rented between 300 and 600 acres of cropland. Marty's father died in 1987. From childhood, Marty had dreamed of farming. With little opportunity to farm in Ohio, he traveled, looking for other places that might offer a better chance. After marrying Mindy in 1986 he attended a two-year college in Idaho, where he studied crops and soils. After that he worked as an assistant manager of an alfalfa-seed ranch in Washington state. But he had aspirations of starting out on his own in Nebraska, which he had visited in 1982.

"I liked it because it was great big. Everything was a quarter section [160 acres, or one-quarter of a square mile]. I really fell in love with Nebraska," he recalls. Eventually, Marty got a chance to work for a John Deere farm machinery dealership in Albion, Nebraska. He had experience working as a mechanic and repairing machinery. There were no openings for mechanics, so Marty got started by working behind the parts counter.

"It was wonderful because I got to know many of the farmers in the area," he says. One of the farmers he met owned the farm across the road from the rural house that Marty and Mindy were renting. When the farmer learned of Marty's interest in farming, he agreed to rent Marty some buildings across the road where Marty started farrowing sows. The farmer helped Marty out by providing him the sows and allowing Marty to pay for them after he sold his first pig crop. Marty borrowed enough to buy feed for the hogs and to buy a boar.

At about that time Marty heard an announcement on the radio about a Land Link meeting for young people interested in farming. After attending the meeting, Marty and Mindy signed up for Allen Prosch's classes, a series of workshops dubbed "Starting Out Right." It was a small group of young people interested in farming. The Hitchcocks enjoyed trading ideas.

In the workshops they learned of Del Akerlund's interest in taking on a young farmer to continue his organic farming. Del interviewed the young couple in December 1991. In February of 1992, Marty and Mindy decided to move to the Akerlund farm, where Marty started working in March for a monthly salary.

Marty moved his 11 sows and a boar to the Akerlund farm, but the rent that Del charged for the buildings and the use of a grinder for the hog feed was more expensive than the arrangement in Albion. On the other hand, Del had to provide a feed mixer and one building because he had not been raising hogs on his farm. Prosch helped mediate between the two.

Otherwise, though, "Del and I got along fairly well," Marty recalls. But as the year went on, Marty felt that he had little influence on the decisions made on the farm. And the hours were longer than he and Mindy had expected, often running into 14-hour days in the summer, even though Del also employed two part-time laborers.

"When I grew up, we would work hard when we worked and then we played hard," Marty says. "I always associated agriculture with owning your own time—but I wasn't my own boss at the Akerlunds and that bothered me. I couldn't have gone on. I would have had ulcers."

Mindy, who didn't grow up on a farm, recalls that "we didn't think it would be easy," but still she was surprised by just how hard Marty had to work. "I hated to see him get up and be tired and have to put in such a long day." To spend more time with him "sometimes I'd put the kids to bed and would go out and ride with Marty in the tractor," she recalls.

Marty, too, felt that he wasn't with his family enough and that his two preschool children were growing up too quickly. So, after harvest, Marty told Del of his plans to go back to college and the two ended the business relationship amicably.

"Marty was a good farmer, but he had other things going not his way," Del Akerlund says, adding that Mindy didn't seem happy. "If the wife is not satisfied, that's half of the family getting along." (Later, Marty said that his wife's feelings reflected sympathy for his own frustrations. "She loved the area and she cried when we left," he said.)

Today, Marty works as a mechanic and is part owner of a diesel repair shop in Provo, Utah. He still works hard, putting in 10-to-11-hour days, sometimes 14 hours, but at least he has time to take his family—now three children—to the mountains on weekends. Business has tripled since he entered into a partnership with an older mechanic. He recalls that in the Starting Out Right workshops in Nebraska, Allen Prosch had said some people will decide farming's not for them. "I never dreamed it would be me."

Somewhat wistfully Marty remembers how he waited in anticipation for the farm magazines to arrive when he was growing up in Ohio. But lately he has put farming out of his mind. "I haven't watched the markets for a couple of years now," he says. The Akerlunds admit to being surprised that their experience with the Hitchcocks didn't work out.

"You can't please everyone," Del muses. Long hours are inevitable during certain times of the year for farmers, especially those who don't use farm chemicals. "If you're not there on time, you just don't make it that year." Despite their disappointment that not every situation in land linking programs can always work, the Akerlunds remain strong supporters of such efforts. "We need some incentives to get young people into farming," Del says.

Out of 47 matches Land Link has made, only a handful have not worked out, Allen Prosch says. Even those seemed to end without any rancor. "I don't feel like Marty and Mindy constitute a failure in the process," he says. They learned from their experience at the Akerlund farm and they were able to make a decision about their future from that.

Part of Prosch's workshop had covered "strategic management," which involves setting goals and learning how to measure progress against those goals, and also how to judge your own strengths and weaknesses. "That's essentially what Marty and Mindy did," Prosch says. They were farming and their hog business had done well. At the end of their stay on the Akerlund farm they owned $3,000 worth of breeding stock. But they hadn't reached all of their personal goals.

Prosch also worked with the next young family that's now working on the Akerlunds, so he knows all three families. "They're all really great people," he says.

The second young family, Tim and Denise Hendrix and their preschool daughters, Jordan and Hanna, seem happy at the Akerlund farm. And the Akerlunds seem charmed by the good-natured young man from North Carolina who stopped by one June evening with a handful of summer squash from the Hendrix garden.

Sitting across from Tim at his dining room table, Del smiles as he recalls Tim's arrival in September of 1993. "We didn't even hardly say anything to each other and the first thing I knew, he was out cutting hay."

"It's so nice to have somebody who knows that sometimes you have to work extra hours on the farm," Merle adds.

And on an organic farm, especially, the hours can be long. Weed control depends on timely cultivation. If you wait one day too long, and a rain delays that cultivation, the weeds will grow too much to be controlled by the time you have a chance to cultivate again. So Tim put in 16-hour days several times in his first year on the farm. One day in June of 1994, "I got done laying by [finishing cultivating] and it rained that night," Tim says. It wasn't the first time his field work had been finished just in time.

Yet, in spite of that hard work, Tim says that "I've spent more time with my family since I've been here."

That might not seem like a credible statement, until you know more about Tim's background. Before moving to Nebraska he drove 11 miles one-way from his home near Moxville, North Carolina, to milk 75 leased cows twice a day. And, to make that work financially, he had to drive a milk truck as well. An older farmer had talked Tim into taking over the lease for the cows in 1989. Tim rented farmland to raise silage, but got caught in two years of drought. Finally, the Hendrix family sold out, able to pay off their PCA loan but left with little else. "We fought it for five years as hard as we could go, and at the end of the five years, we was at the same place we started. The money me and my wife had saved to put into it, it was all gone," Tim recalls.

At 32, Tim already has a lifetime of experience trying to get started in farming. "I started with nothing—bare necessities," he says. "My family farmed until I was six or seven years old." Then his father and grandfather sold out. His father went to work mowing highways for the state and his grandfather kept a small truck farm and ran a barbecue restaurant. At 15, Tim started working part-time for a dairy farmer until he finished high school. Then Tim drove a truck for about six years, before starting to work in town for a feed mill. There he came across several farming opportunities. First, he bought hogs from a farmer who had developed diabetes. He rented land to raise the feed and bought "some old junky equipment to start out with." Later, a dairy farmer talked him into buying into a partnership to milk cows. Tim abandoned that because his partner usually showed up for work after Tim had already finished milking. After that he took over the 75-cow herd from another dairy farmer, who was going through a divorce.

In retrospect, Tim believes that he got involved in some bad business deals. "I didn't stand up for myself and I got taken advantage of a few times," he says. "This trying to get started on your own is just too hard. You beat your head against the wall."

In contrast, things seem to be going well at the Akerlund farm. Tim's wife, Denise, works as an administrative assistant at an Omaha telemarketing firm. Del says "I'm paying Tim well because he's worth it." And Tim was renting an additional 160 acres of cropland in 1994, with Del getting a share of the crop in return for Tim's use of Del's machinery.

Finding that opportunity wasn't easy. After Tim saw Land Link mentioned on a television news magazine, it took ten phone calls to Nebraska before he reached the Center for Rural Affairs. He and Denise first visited Nebraska in October of 1992. "Allen showed me, I don't know how many farms. We stayed out here about a week, " he says. They liked Nebraska but found no farm that seemed like a good prospect for them. Some were for sale, and the cheapest down payment was $50,000, which the Hendrix family couldn't afford. Other farms seemed too run down, with buildings that had been out of use for three or four years. They met the Akerlunds on a second trip to the state in August of 1993.

When asked how Del and Tim split up the work, Merle interjects, laughing, "Tim does most of it." Besides doing much of the work on the farm, Tim also negotiated the lease for the extra 160 acres he farmed on his own. Del sometimes has him negotiate with buyers of the farm's organic grains. To Tim, the idea of negotiating a price for farm products is a novelty. In North Carolina he had to wait 45 days for each milk check set at market prices. "This is something we can set the price on. We don't just take something to town," Tim says.

After about two years, the working arrangement between Tim Hendrix and Del Akerlund seems to be off to a good start. Tim does get time off, including nine days that he took to visit his family in North Carolina. The two men still don't have any formal long-term contract on how control of the farm will be transferred to Tim. But Del talks as though it can be a long-term business relationship. "I'd say he's here for a long time. He's going to make it," he says. Tim adds, "This is a good way for a young person to get started."

Putting the relationship in writing and having some type of written transfer agreement, such as a buy/sell agreement that transfers the estate at the death of the older farmer, is important, Prosch says. But getting to that point can take a few years, even though Prosch, Baker and others who run state linking programs strongly encourage written agreements.

"The idea is that, if you can't get it in writing, either one or both parties don't have a clear understanding of what they're trying to do," Prosch says.

Helping farmers to reach that understanding, to have realistic goals on the part of both the older and younger farmers, is the essence of Prosch's work. He no longer conducts the Starting Out Right workshops. They've been incorporated into classes given by two-year community colleges in Nebraska. Farmers Home Administration in that state requires its young borrowers to take the classes. And they're open to any other young person interested in farming. But Prosch and his co-worker in the Land Link project, Joy Johnson, work individually with young farmers. And they also hold "alternative estate planning workshops" for older farmers in Nebraska and western Iowa.

After an older or younger farmer signs up to be in the Land Link database, its staff meets with each person or talks to them by telephone to help them focus on their goals. Prosch might spend three or four days total over a period of several months working with each person, taking up to an hour at a time with numerous follow-up phone calls. Land Link charges $15 an hour for its services, although Prosch winds up donating a lot of his time.

Before any formal match takes place, "there's a process that everyone goes through and it takes six months to a year. It's a process of people coming to grips with what they're going to do," he says. "It's more difficult than a son coming home to work on the farm." That's because most parents have a sense of obligation to help a child get started. They're not likely to have that kind of emotional commitment to work with a stranger. So clear goals are important at the outset.

Often, when Prosch talks to older or younger farmers, the goals aren't realistic. An older farm couple might expect a young couple to work 18 hours a day so that they can retire to Florida and collect a check, Prosch says. "We try to walk people through this process. We'll ask, 'Where are your children now?' and they'll say, 'Oh, well, they moved off the farm when they were 18 because we worked them to death.' "

"You get the same thing from the younger folks," Prosch adds. Some expect to simply move onto an older couple's farm and take over, running it completely on their own. That's not realistic either, at least at the very beginning. "You're talking about the older farmer's assets. You're living in their checkbook, almost," Prosch says.

Ideally, the older couple will start this process in their 50s, not the year they retire. That gives the young family working with them more time to gain experience and to build up enough savings and equity to buy out the older farmer gradually. But, because the matching programs are fairly new, that's not always possible.

After Prosch feels that the farmers are prepared and are being realistic, he starts the matching process. In 1994, Land Link had 1,300 beginning farmers and 150 older farmers in its database. (Most are from Nebraska and Iowa, but Land Link takes names from anywhere in the country. Land Link once helped find a beginning farmer for a Florida orange-grove owner.) Out of that number of farmers, only about 75 older farmers "are really active" in the program, Prosch says. And only about 300 to 400 younger farmers are mature enough and ready to start farming. "There are a lot of daydreamers" among the younger names in the database.

Prosch then sends the names of older farmers to young farmers who are interested in the type of farming practiced by that older farmer. In some cases, one older farmer could potentially be overwhelmed with as many as 50 phone calls. But in practice, only 10 or 12 young farmers will actually call the older farmer. And some may wait five or six months before calling. The older farmer usually winnows the list down to three or four candidates that he interviews.

Prosch has found from experience that it doesn't work to have someone else match up the older and younger farmers. By now, he has abandoned any active role in the final decision about who goes to each farm. "We found that it's much better to let them do their own selecting," he says.

Sometimes the older farmer hires the young person to work and Prosch never hears from them again. In other cases, the retiring farmer has a hard time choosing between several candidates and may give Land Link a call. "This is where we get back into the real meat and potatoes of how much help they're really seeking," he says. In one case, Prosch helped write a draft of a work-in agreement. Neither the older nor younger farmer felt comfortable signing it, but they refer to it and scribble in changes each year "and they get along fine." The agreement covers a ten-year transfer of the older farmer's swine herd to the younger farmer. In one year the young person had a poor return, so he didn't expect a big bonus. The next year, he did well and the older farmer gave him a bigger bonus than the agreement called for. At the end of ten years, the transfer will be complete "without very much pain or tax consequences for the owner," Prosch says.

Of the 40 percent of matches made by Land Link without any written agreement, the older farmer is hiring the younger farmer as an employee. There is usually an unwritten verbal agreement that the younger farmer will have more opportunities after a trial period. "The older farmer might say, 'I'm thinking about turning this over but I'd like to see you work a year here first,'" Prosch says. A trial period of six months to a year is good for the younger farmer, too, he says. It gives him or her a chance to see how well the farm is doing financially. You don't necessarily have to look at the books to know. If the farmer can't afford to buy replacement parts for his machinery and has to hold things together with baling wire, it's not a good sign.

Exactly how closely the older and younger farmers work together will depend on each one's personality. If the younger farmer doesn't want the older one looking over his shoulder every day, that should be discussed.

Earl Wright of Kansas Farm Link urges older and younger farmers alike to carefully negotiate any potential working relationship. Photo by Michael Malone, courtesy of *Successful Farming.*

But, again, there is no formula for a good relationship. "One guy's independence is the next guy saying, 'The boss never told me what he wanted to get done,' " Prosch says.

So far, Prosch has seen no real failures in the Land Link program. Even the matches that haven't worked should be viewed as valuable experiences, he believes. "The only situation I would feel bad about is the guy who starts out at 25 and by the time he's 40 he gets kicked off because the older farmer gave the land to the church," he says.

But potential for exploitation in these programs does exist without careful negotiation on both sides, says Earl Wright, who runs Kansas Farm Link for that state's agriculture department.

Wright, who has a background in hog farming, teaching and running the state's farmer-lender mediation service, starts first by looking at the financial plans for transferring an older person's farm business to a younger family.

"What you need is a business plan, a financial plan, to see if there is enough there that will satisfy retirement needs as well as family living needs," he says. "Off-farm income is many times a key element in that. But ultimately the goal is: Let's make a living from the farming operation."

If it looks as though the established farm is profitable and big enough to meet the needs of two families, the next step is to work out an agreement that's fair to both sides. Older farmers shouldn't sell their assets for less than fair market value. "No one's going to ask you to do that" through state linking programs, he says. Nor should young people work hard without getting fair compensation. "You need to be assertive," Wright tells younger farmers.

Above all, though, "there needs to disclosure on both sides," he says. "The landowner and beginning farmer prospect need to be open. I can almost guarantee that if somebody's got a hidden agenda on either side, it may work for a year, maybe two or even three years, but it probably is going to come apart when that hidden agenda finally comes out."

If the younger and older families are negotiating a long-term agreement, a lease or selling the farm on a contract, Wright urges extra caution. "If you're uncomfortable with an agreement, on either side, I think you should just back out of it. That's a pretty harsh statement, but that's a choice. You may just have to walk away from it. There are opportunities out there. They're not easy to find and they're hard to execute. But getting into a bad deal is much worse than looking for another deal."

Despite that caution, Wright urges older farmers who do not have children interested in farming to consider state matching programs. Like the Center for Rural Affairs' Land Link program, Kansas Farm Link works with farmers outside of the state.

Precisely because of the work of people like Earl Wright and Allen Prosch to encourage good working relationships between older and younger farm families, the new state linking programs may well be the most exciting development in the decade of the 1990s for helping the transition to a new generation of farmers. They provide a third party that can act as a mediator when the younger and older farmer run into problems.

The services are relatively inexpensive. Prosch estimates that the Center's Land Link, which has two people spending part of their time on the project, costs about $50,000 a year. Some states could get by with one half-time staff person, he says. But they'll have to be dedicated. "I've gotten calls at midnight waking me out of a sound sleep," Prosch says. Sometimes the caller really doesn't have a problem. "They call just to be encouraged—and that's part of the process."

And, as the experience of Marty Hitchcock shows, some opportunities to farm come outside of any formal program. In retrospect, Marty thinks that he might have done better to continue working for the John Deere dealership in Albion, Nebraska, and raise hogs with the help of an older farmer. The farmer ran a big operation with several sons, but still had helped several young farmers get started in hog production, Marty recalls.

Prosch agrees that there's nothing really unique about such arrangements. "There always were individuals who just went out and did this," he says.

But, as more farmers near retirement age, more matches will be needed—either informal ones or those made through state programs. So far, a relatively small number of matches have been made through the formal process. In Iowa, Farm-On had made 19 matches between August of 1992 and summer of 1994. But Farm-On's John Baker has talked to many more older farmers interested in farm transfers. And Iowa State University's farm management specialists are getting an increasing number of questions about exactly how such transfers can be made, he says. Baker heads an informal group of state programs, the National Family Farm/Ranch Transition Network. He believes that in spite of their newness, the matching programs have already been effective.

"I think they have made tremendous inroads in the farm community, of bringing out the need to look at alternatives," he says. "Obviously, it's a big task and there's a lot of work to do. But I see the beginnings of those efforts to do that work. We're all getting better at it. I'd give them an A, overall."

CHAPTER 7

Federal Help: The Promise of Beginning Farmer Down Payment Loans

Greg Quiring, 37, had already been running a small cow herd on about 60 acres of rented land near York, Nebraska, when the opportunity to buy part of that pasture came up.

In fact, after a six-year part-time career in the cow-calf business, the chance to farm was thrust upon him. The cropland on a 157-acre farm where Greg rented about 30 acres of pasture for his cows had been rented out to a local crop farmer. But now the entire farm was up for sale. Greg and his wife, Nancy, and their three children also lived in the farmhouse on the land that was on the block. In order to be assured of keeping part of his rented pasture and their home, they might have to make an offer on the land.

Fortunately, Greg's loan officer at First National Bank of York, Gus Brown, had heard about a new program that might make buying the farm possible for the young couple. Greg, who grew up on a small farm at nearby Henderson, puts up and services irrigation equipment used by area farmers. Nancy is business manager at a medical clinic in York.

The new program, sometimes called Beginning Farmer Down Payment Loans, is offered by the Farmers Home Administration, the lending arm of the U.S. Department of Agriculture. It started operating in the fall of 1993 and is part of the agency's older Direct Farm Ownership loan program. (When this book was going to press, the future of the agency's direct loans appeared in doubt. If they continue, they will be run by a new USDA agency, the consolidated Farm Service Agency, whose main job will be to run crop price support programs.)

In recent years, loans made directly by the federal government to farmers had been a shrinking program. Under efforts by presidents Reagan and Bush to privatize farm lending, the federal government had been putting more money into providing guarantees to banks who lend to farmers. In the first two years of the Clinton Administration, the emphasis on loan guarantees continued. Although the guarantees were also aimed at farmers with modest resources, they were less helpful to young farmers than older direct loan programs. Even with a federal guarantee, a bank usually requires at least 25 percent down on a land loan. That's better than the 40-to-50 percent normally required as a down payment, but still out of reach for young farmers. Farmers Home's older direct loan programs required no down payment of young farmers, but funds for those direct loans were also being squeezed by the new emphasis on guarantees and by budget cutting.

Under the new beginning farmer loans, a young farmer must put up a down payment for 10 percent of the farm's value. Farmers Home Administration provides 30 percent for the rest of the down payment. Then the remaining 60 percent is financed by a commercial lender, such as the bank in York, or a landowner willing to carry the debt through a contract. For Greg's loan, the bank lowered its own interest rate by using the Nebraska Investment Finance Authority, an aggie bond program.

Greg and Nancy had to put $14,000 down on the 157-acre farm, which they bought for $140,000. They borrowed from their life insurance for the down payment. Farmers Home Administration financed 30 percent at a 4 percent interest rate. First National Bank of York financed the remaining 60 percent at a little more than 8 percent interest.

Greg thinks that other young farmers interested in this program will be helped if they have a long-term relationship with a bank, since the new program requires the participation of a commercial lender or landowner. Greg and Nancy had been borrowing at the York bank for more than a decade for consumer loans and for his investment in cattle.

"This is a whole new program and I'm not sure anyone knows how to handle it," he says. "Now everybody has to kind of share you with everybody else. Now you've got more than one boss to work with."

Under the new beginning farmer loan program, Farmers Home Administration can also make direct operating loans to a young farmer for up to ten years. But Greg is getting his operating loan at his York bank.

Greg was cautious about his venture into crop farming. He chose not to buy any machinery the first year. Instead, he hired another farmer to "custom plant and custom harvest" for a fee.

"The biggest thing I'm trying to do the first year is to just survive and see what the land produces," he says. If all goes well, he'll start buying machinery after that.

In theory, the down payment loans should allow Farmers Home Administration to help three times as many young farmers as it did with direct loans. But in practice, the program didn't start out that way. Delayed for nearly a year after President Bush signed the new law in 1992, the program got off to a slow start. Nebraska was one of only a half-dozen states to make the loans when regulations for the program took effect in the fall of 1993. Nationwide, only $26 million out of the USDA's total $65 billion budget went to the new beginning farmer loan program.

Nebraska Farmers Home Administration director Stan Foster, a farmer and former bank loan officer, was one of the most enthusiastic promoters of the program. Even before his agency could make the new loans he trained his staff and met with farm group leaders and bankers to explain it.

At The First National Conference for Beginning Farmers and Ranchers in Omaha in March 1994, Foster told young would-be farmers and ranchers that they would have to seek out the new loans.

"You need to be aggressive," he said. "You've got to take the initiative to contact your county FmHA office. Find out where that is and who's in there and go in and introduce yourself. Schedule an appointment so you can discuss what the beginning farmer programs are. Then you need to gather data for your own records—a financial statement, a cash flow, a history of your production and a history of your financing. ... The more complete your application, the faster we can process it. ... You need to be in contact with a lender. If you're going to get a land contract, you need to contact a seller."

Foster said the typical beginning farmer borrower in Nebraska is organized, has set goals, and is "pleasant to work with." Most have some experience farming or working on a farm. "Does off-farm income disqualify you? It does not," he added. "It depends on what level of income you have, but the beginning farmers we've been working with have off-farm income, or the spouse does."

The loans offer several attractive advantages. The bank or contract seller's portion for 60 percent of the farm's value is amortized over 30 years. A bank's portion of the loan can be guaranteed by Farmers Home Administration. Or, in states with an aggie bond program, the bank or landowner can use the bonds to avoid paying federal income tax on the interest paid by the beginning farmer. That lowers the interest charged to the young borrower. The bank takes the first lien on the property and Farmers Home Administration takes a second lien. When the program began, Farmers Home's interest rate on its 30 percent portion of the loan was 4 percent.

Other characteristics of the loans make them tougher for some borrowers. The 30 percent "down payment" loan made by Farmers Home must be paid back in ten years. "This program is targeted to build equity fast," Foster explained. "Not everyone has the cash flow to make this program work. The first thing you'll discover about it, is it takes more money each year to make the payments because we're on a shorter fuse. ... At the end of ten years you've put in 10 percent, paid off 30 percent and in the meantime, you've paid on your traditional loan for the remaining 60 percent, so you're right at 50 percent equity. Then you're eligible to graduate and go on to traditional credit."

Like all Farmers Home Administration programs, the beginning farmer loans are designed to have a limited clientele. To be eligible for a loan, the young farmer can't be too well off, yet his or her proposed farming business has to be financially viable. Farmers Home has a basic list of criteria for all of its borrowers. A local county supervisor decides if each loan applicant meets those seven characteristics—being of legal age, having U.S. citizenship, having educational experience to farm, having a good credit history (if you've gotten small loans to buy livestock or feed), being unable to get credit anywhere else (in this case for the bigger land loan), being able to show that you will own and operate the farm you're buying— and you must provide honest information.

Beyond that, the beginning farmer program has a few more rules to meet. A good summary was prepared by Oklahoma State University agricultural economist Ross Love and extension assistant Lori Shipman. First, they wrote, the borrower has to meet the government's definition of a beginning farmer. That's one who:

• Has ten years or less experience running a farm or ranch. (It doesn't have to be ten consecutive years, Foster adds.)

- Has "materially or substantially participated in the day-to-day labor and/or management" of the farm or ranch, enough that the business would suffer a lot without that participation.
- If he or she already owns land, it must be less than 15 percent of the average farm size in that county. (In Foster's Lancaster County, Nebraska, that's less than 56 acres.)

Second, the proposed agricultural operation has to be considered financially viable by Farmers Home Administration. That includes evidence that:

- The farmer plans to reach a financial position that allows him or her to operate without more credit or bank loan guarantees from Farmers Home Administration.
- The farm or ranch will be able to make loan payments on time.
- The farm will provide for any living expenses that are needed in addition to the family's off-farm income.
- The farm generates enough income to replace capital purchases such as machinery as it wears out.
- The farm or ranch has "continuing financial vitality."

It's important to know, too, that the loans are relatively small, limited to enough credit to buy $250,000 worth of land. That can be a big debt for a young family with few assets. But it's too little to buy enough land to support a family on grain production in most areas of the Corn Belt. To support a family, the beginning farmers in this loan program will obviously have to have livestock on the farm, an off-farm job, or be able to rent much more land at favorable rates—or maybe all of those sources of extra income.

Even though young farmers must show that they have some experience in order to get the loans, they are also required to take "borrower training."

"In Nebraska it's set up with our community colleges," Foster says. "It's a very good course. Two years cost $600." The county supervisors can waive that requirement if they decide the young farmer has already had enough education.

Beginning farmers don't need to own everything that they plan to use in their farming business. But they must show in their financial plans that they have access to the buildings, machinery or other equipment they'll need.

The new down payment loans aren't the only programs available to beginning farmers. The agency has a small amount of funds available for direct loans to beginning farmers with 40-year amortization. It will finance up to 100 percent of the purchase. When Foster talked to beginning farmers and ranchers in Omaha, the interest rate was 6.5 percent. For farmers with "limited resources," it was 5 percent.

Farmers Home Administration also has several type of direct operating loans that beginning farmers can use for buying equipment or to pay annual operating expenses. The rates in the spring of 1994 were at 5 percent or higher. A special rate was available for beginning farmers who had farmed five years or less.

Besides making loans, the agency can do other things to help beginning farmers, says Chris Beyerhelm, chief of farmer programs for the agency in Iowa. For example, the agency allows debt it is owed by farmers who are quitting farming to be transferred directly to a young farmer. That can ease the pain for those older farmers if they know that the end of their career on that farm will help another family farmer get started, he says. On his own, Beyerhelm has also encouraged local Farmers Home Administration staff to organize county-level groups that might help match older farmers with beginning farmers.

Because some of the older loan programs aren't quite so difficult to pay off in the early years, Beyerhelm was among many professionals in the agency who had some early misgivings about the new down payment loans for beginning farmers.

With the new loans, "if you've got to pay 8.5 percent to 9 percent interest on that 60 percent [borrowed from a bank or landowner] and accelerate the repayment on our 30 percent, it can be hard to cash flow," Beyerhelm told *Successful Farming* magazine shortly after the program went into effect.

Like Nebraska, Iowa has an aggie bond program that Beyerhelm and his staff were beginning to work with. When Iowa banks used aggie bonds, some were able to lower the interest rate down to 6 percent. With that cooperation, the down payment loan program "really works slick," Beyerhelm says.

Unfortunately, less than half of all states have aggie bond programs, so the down payment loans will be harder to make cash flow in the remaining states, unless a young farmer can find a landowner willing to make a contract sale at favorable interest rates.

Even in some states with aggie bond programs, Farmers Home Administration officials weren't as aggressive at using the down payment loans as Nebraska and Iowa.

Nancy Thompson, the Center for Rural Affairs attorney who helped get the down payment loan program through Congress, says it seems more complicated to agency staff that are used to making loans on their own or with guarantees to one commercial lender. The down payment loans can involve a bank or contract seller, a state aggie bond program, and Farmers Home Administration. "This is the kind of program that does take more effort to make it work because it does require coordination of a lot of different actors," she says.

Finally, even the Farmers Home Administration officials who seem to care about helping young farmers could hardly be blamed for lack of enthusiasm for a new program. Under Agriculture Secretary Mike Espy's plan to reorganize the USDA, Farmers Home Administration has been split up, with its farm lending programs folded into a new consolidated Farm Service Agency that will handle the USDA's much larger price support programs. Farmers Home's rural housing and rural development loans would shift into a new Rural Economic and Community Development agency.

Not only was there uncertainty about how much emphasis the agency's farm loans would get in a larger agency created to run most farm programs, but Farmers Home Administration's budget for direct loans for farm ownership and operating expenses was cut by Congress in its budget for the 1995 fiscal year.

Congress cut the authority for making operating loans from $700 million in 1994 to $500 million in 1995. It froze the agency's lending authority for the farm ownership loans at $78 million—but it cut the budget for subsidizing those ownership loans from $13 million in 1994 to $11 million in 1995. With interest rates on the rise, it appeared unlikely that the subsidy really would be enough to make another $78 million in farm ownership loans. The good news for young farmers is that the beginning farmer law requires a larger share of those funds to be devoted to down payment and young farmer operating loans each year.

Farmers Home Administration's loans were just one of several worthwhile programs cut. The Soil Conservation Service saw its budget trimmed by $300 million.

Congress was locked into cutting just a few programs because others in the USDA budget, including food stamps and farm commodity subsidies are "entitlements" that can be changed only once every five years when the federal "Farm Bill" is written (unless Congress decides to amend the Farm Bill). The net result was that a few "discretionary" programs like Farmers Home Administration took cuts of 10 percent. Unless farm programs are changed drastically, these cuts are only likely to get worse because Congress has passed deficit-cutting legislation that requires even greater cuts in the federal budget in future years.

In 1995, at least, the result of the budget cutting was that, once again, the established farmers who tend to benefit more from price support programs weren't hit as hard by budget cuts as beginning farmers, who don't have a strong voice in Washington.

The following year, when the Clinton Administration submitted its budget for 1996, it proposed increasing the amount of money spent on direct loans by Farmers Home Administration. Sustainable Agriculture Coalition lobbyist Ferd Hoefner took that as a good sign for the outlook for beginning farmer down payment loans, which come from those direct loan funds. But Congress may have a different view. Senator Richard Lugar (R.-Ind.), the chairman of the Senate Agriculture Committee, told *Successful Farming* that using the federal lending agency to guarantee bank loans might be better policy, given the losses the agency has incurred from its existing loans.

When the Republican Party took control of Congress in the elections of 1994, the party's commitment to balancing the federal budget put the future of all Farmers Home lending programs in doubt. Even before the election, some Washington insiders expected the agency to eventually die. Still, its program of guaranteeing farm loans is popular with banks and that might be continued. Whether or not those loan guarantees are used more actively to help beginning farmers likely will depend on the interest of the private banking community.

In the meantime, as long as the beginning farmer loan program still exists, the budget pressures in Washington are a good reason to follow Stan Foster's advice to "be aggressive ... to take the initiative to contact your county FmHA office."

CHAPTER 8

State "Aggie Bond" Programs: Tax Breaks for Agricultural Development

Bill Greiner, now retired, ran Iowa's Agricultural Development Authority, one of the nation's most-used aggie bond programs. Photo by Michael Malone, *courtesy of Successful Farming.*

From the deck of his spacious ranch house Al Wiebelhaus has a fine view of the trees and rolling hills of western Omaha's residential neighborhoods. The retired manager of a Safeway meat department did well in his career. He has done even better in retirement, profitably investing in real estate. But he hasn't forgotten his farm roots. So, when a nephew, Tim Pick, told him of a 248-acre farm for sale in Cedar County, Nebraska, Wiebelhaus was interested.

"Tim said, 'You buy that and I'll take good care of it for you,'" Wiebelhaus recalls. So he did, paying $215,000 and taking possession of the land in 1991. Pick and his brother-in-law rented the farm from Wiebelhaus.

Wiebelhaus might have been a landlord for a long time, but one day he was reading a real estate magazine in a broker's office and he learned about an agricultural lending program run by the Nebraska Investment Finance Authority. The authority issues bonds that are exempt from federal income taxes to finance industrial development. This type of bond has been used for years to finance new factories. They were called industrial development revenue bonds until federal tax reform legislation changed the name to "private activity bonds" in 1986. Income from the bonds is exempt from federal taxes.

Iowa was the first state in the nation to use revenue bonds to finance loans for beginning farmers, starting in 1981. Nebraska followed with the NIFA program not long after that. Basically, if Pick met NIFA's eligibility requirements, NIFA would swap an "aggie bond" for a sales contract Wiebelhaus wrote for Pick. NIFA would then assign the sales contract to Wiebelhaus as security for the NIFA bond. In practice, it would be like any other installment sale contract. Pick would have to make annual payments to his uncle for 30 years. But Wiebelhaus wouldn't have to pay federal income taxes on the interest he receives from his nephew.

"I thought maybe that wouldn't be a bad idea to sell it on a NIFA bond, but we both didn't think he could afford it on NIFA alone," Wiebelhaus recalls. But, when the new beginning farmer loans were offered by the federal Farmers Home Administration in late 1993, that made the aggie bond contract work better for both buyer and seller. Pick borrowed from Farmers Home to pay 30 percent of the $215,000 that his uncle charged for the land. Pick put up another 10 percent. And his uncle carried the rest of the debt through an installment loan contract for $129,500. Wiebelhaus got a bigger down payment—40 percent—than is common practice on sales contracts, which usually require 10 to 20 percent down. And he felt more secure knowing that the federal government had a second lien on the property and wouldn't let it be sold cheaply if Pick couldn't make the payments. Pick got a 4 percent loan from Farmers Home on its "down payment loan." And Wiebelhaus wrote the contract so that Pick made smaller payments on the principal during the first 10 years, when the young farmer has to repay the Farmers Home down payment loan.

Tying the Farmers Home loans into an aggie bond loan is a new wrinkle on the older aggie bond program and Wiebelhaus considers the new arrangement a "big sweet deal" for both buyer and seller.

But even if the aggie bond program is used alone, it's a very good deal for the seller and helpful for the borrower. In the sale Wiebelhaus made to Pick, the landowner charged a 7.7 percent interest rate to his nephew. That's slightly below the rate of 8 to 10 percent commonly charged on land contracts at the time if they weren't backed by aggie bonds. It's a little higher than some 6 percent loans made when interest rates were extremely low in 1994, but Wiebelhaus made other favorable adjustments to the contract, such as accepting lower payments while Pick pays off Farmers Home Administration, and asking no more than his purchase price for the property.

For Wiebelhaus, the 7.7 percent yield on the bond is a very good investment compared to more common tax-free investments such as municipal bonds, which were yielding about 6 percent at the time. And, if Wiebelhaus bought bonds that weren't tax-free, he would have to get a yield of about 10 percent to have an equivalent investment. The aggie bonds are also exempt from state income taxes in Nebraska. Aggie bonds have slightly more risk than investments like treasury bonds, though. NIFA and similar bonding authorities in other states don't guarantee the bonds. That means that if the young farmer who buys the land through an aggie bond land contract can't make the payments and defaults, the seller would get the land back. In the 1980s, when farmland values crashed, many older farmers who sold land through private contracts (not tied to aggie bonds) did get their land back. They suddenly found that it was worth about half as much as when they sold it, and their retirement income was reduced considerably. That risk still exists, although most observers of the farmland market think such a crash in land prices is unlikely soon. The farm debt crisis of the 1980s was the worst in more than a half-century. The land market today is considered much more stable, rising at about the rate of inflation in the 1990s. Prices began to rise at a slightly higher pace in 1994, partly because older rural residents could get a better return on land than on savings deposits and were in the land market. But as inflation started to pick up again, that advantage might prove to be short-lived and savings and other investments might appear more attractive than land.

Aggie bonds have some advantages over conventional contract sales of land. Even though bonding authorities don't guarantee the loans, their requirements do make contract sales more businesslike. Wiebelhaus had to go to several board meetings of NIFA to get the loan approved. The board makes the decision based on whether the buyer meets eligibility requirements under federal tax laws. Those requirements will be described in more detail later, but basically, the beginning farmer can't have too much net worth or already own more than a small amount of farmland. (The seller has to make his own determination as to whether the buyer is creditworthy, however.)

NIFA has a bank act as an escrow agent. For the Wiebelhaus contract sale to Pick, it's First Dakota National Bank in Yankton, South Dakota, which is just across the Missouri River from Tim and Kathy Pick's farm near Hartington, Nebraska. The Picks make their payment on the land contract to an escrow agent at the bank. The agent reports the payment to the Nebraska Investment Finance Authority and puts the money into Wiebelhaus's savings account at the same bank.

Wiebelhaus keeps the bond in his briefcase with other real estate investment records. After a minute of leafing through his papers he pulls out several sheets of tan paper stapled together. It wasn't an embossed document with a fancy border, just several sheets of typed legalese. "You'd think you'd see a bond looking like a bond," Wiebelhaus says. "It isn't. It's just pieces of paper."

Wiebelhaus found the process of getting the bond issued to be fairly simple and the bank proved helpful. In Nebraska, a land seller can get help on using aggie bonds at any bank. Now that the Farmers Home Administration beginning farmer loans make aggie bonds even more attractive, he's surprised that more retiring farmers don't use aggie bonds.

If a young farmer gets a Farmers Home down payment loan, he'll pay the older contract seller 40 percent of his farm's value up front, Wiebelhaus says, "which means 40 percent of the money comes to him. He can buy a house in town and the rest of this is coming in tax free—spread out over 30 years." That way, the older farmer is likely to avoid most, if not all, income taxes on the sale of his land.

The biggest disadvantage to aggie bond programs is that they aren't available in most states. In 1994, only about a dozen states, mainly in the Midwestern Corn Belt, had bonding authorities to use the federal program. (See Appendix for a complete list.)

Aggie bonds can also be used by banks, which get some of the tax breaks that contract sellers get. The banks using aggie bond programs also lower the interest rate to their young borrowers.

In Iowa, for example, aggie bonds have been available to contract sellers only since 1987, says Bill Greiner, who recently retired as the executive director of the successful Iowa Agricultural Development Authority, that state's equivalent to NIFA in Nebraska.

"Banks are still our biggest supporters," Greiner says. "We hear a lot of criticism of banks, but they do want to help people, most of them." It is not just the tax breaks of aggie bonds that motivate banks to use them, Greiner says. "Many of the banks we do business with are rural banks, in small towns of maybe 500 to 1,000 people, and they recognize that beginning farmers or young people on farms are going to be their lifeblood down the road. So they want to help that person either get established or become better established in farming."

One banker who has long had a strong interest in helping young farmers is Jeff Plagge, president of the First National Bank of Waverly, Iowa. Plagge was recently chairman of the American Bankers Executive Agriculture Committee, which represents agricultural lenders who belong to the American Bankers Association. As a leader in the ABA, Plagge has been an outspoken advocate of young farmers. And at his bank in Waverly, and at his previous job as an officer of a Webster City, Iowa, bank, he has worked with many young farmers. Plagge's banks have made aggie bond loans to young farmers to help them buy land.

Partly because of state and federal bank regulations, commercial lenders require higher down payments than a contract seller using aggie bonds, Plagge says. The minimum is usually 25 percent down. Because contract sellers don't have as much overhead as a bank, they can also take a smaller down payment, commonly between 10 and 20 percent, Plagge says.

In some cases, a young farmer has been able to buy a farm through a contract backed by aggie bonds when an aggie bond loan from the bank wasn't feasible, Plagge says. His bank has sometimes advised the buyer and seller to use a contract instead of a bank loan. "We've flat made some recommendations on that."

At the national level, banks have been strong backers of the federal aggie bond laws, which have sunset clauses and periodically have to be renewed by Congress. The American Bankers Association has lobbied for the aggie bond legislation. "When aggie bonds come up in Congress, we make sure we give it a lot of strong support," Plagge says.

Generally, Greiner and others working with young farmers consider aggie bond programs as a midlevel type of assistance to beginning farmers. Aggie bond loans are a better deal for the young farmer than most other commercial loans but they're aren't quite as helpful as a direct loan from Farmers Home Administration. In some cases, a young farmer who is too well-off to qualify for a Farmers Home Loan would be able to use an aggie bond loan. Aggie bond loans have no requirement that the farmer can't be eligible for credit at a commercial bank. Young people usually have to have more equity to successfully use an aggie bond loan than they would need for direct loans from Farmers Home Administration. And the interest rate break isn't quite as good through aggie bonds as through Farmers Home.

Greiner points out that the state authorities for aggie bond programs don't set the interest rates that borrowers charge. That's negotiated between the lender and the borrower. Generally, because of the federal tax breaks, lenders can afford to cut interest rates by about two percentage points. "If rates are running 8 to 8.5 percent conventional, we're going to see about 6 to 6.5 percent," Greiner says. "That's all you're going to get is a reduction in rate. You might get a little longer term."

The lowest rate Greiner has seen was just under 5 percent when interest rates were hitting bottom in 1994. When Greiner made his first loan in 1981, it was at 14 percent, at a time when conventional rates were running between 18 and 20 percent. The lowest down payment he ever saw was 5 percent by a contract seller who said the young couple buying his farm were "the type of people that I want to have my farm. I don't want somebody that owns 1,000 or 2,000 acres taking over my farm," Greiner recalls the seller saying.

Like Farmers Home Administration loans, aggie bond loans are limited in size by federal law. The maximum that anyone can borrow is $250,000. The loans don't have to be used for land, only, although that's the most common purpose. They can be used to buy livestock buildings, machinery and breeding livestock. Out of that total loan limit, only $62,500 can be used for depreciable property. Iowa makes loans for used farm machinery. Nebraska does, too, but it puts other restrictions on those loans. That smaller amount can also be loaned for "used livestock," which, in the case of breeding females is an animal that has given birth. Open or bred heifers and gilts are considered new livestock. Greiner's agency has discouraged purchasing new farm machinery with aggie bond loans. Aggie bond loans can't be used to buy a farm home, but Greiner says his authority often dedicates the down payment to the house.

Another important restriction on the program that's not found in Farmers Home Administration programs is that aggie bond loans cannot be used by a parent to sell land to a son or daughter. Nor can a borrower use aggie bond loans to buy from a spouse, brother or sister, any other ancestor or lineal descendant. Many land sales within families are between parents and children and Congress wasn't willing to open up the federal income tax loophole of aggie bonds that wide. There is no easy, legal way to circumvent that restriction. A landowner can't sell to a third party who would then sell to the children, says Morris Reynolds, deputy director of NIFA. "I've had that question asked many times and I've said that's not possible." If the Internal Revenue Service audited the third party sale, it likely would require that the land not be sold right away to the children, Reynolds says, "maybe for up to five years." (In 1995, Congress was being lobbied to change the law to allow parents to sell land to children using aggie bonds.)

These are the eligibility requirements for using aggie bond loans. Borrowers must:

- Be a first-time farmer who has never owned substantial farmland (defined as 15 percent of the median farm size in his or her county. In Iowa, it ranges from 14 to 48 acres. This, too, might be changed by Congress—to 30 percent of a county's average farm size).
- Be actively engaged in farming or ranching after receiving the loan. Aggie bonds cannot be used by investors.
- Be of majority age.
- Usually have a modest net worth, below $200,000 in Iowa, below $300,000 in Nebraska, below $250,000 in Illinois. Kansas has no net worth requirement.
- Have access to machinery (owned, rented or borrowed), under Iowa's requirements, if the loan is for land. In Iowa, if it's for machinery or livestock, the borrower has to have use of land.
- Have knowledge of the type of farming being financed by bonds.

Greiner says that in Iowa, borrowers can have full-time jobs and get an aggie bond loan. If the loan is for cash grain farming only, he prefers that the borrower have at least a part-time off-farm job. "You cannot take care of your family with the income you're going to get on a grain farming operation," he says. "I've got a grain farming operation and I know firsthand that you can't do it."

Although aggie bond programs are aimed at beginning farmers, they have no age limit and it's not accurate to portray them as only for young farmers. The Iowa program has arranged loans for farmers ranging in age from 18 to 62.

According to Judy Frazier, acting executive director of the Iowa Agricultural Development Authority, the average size aggie bond loan in the 1994 was $112,386. Since the Iowa program began operating in 1981, it had made 1,677 loans for a total of $153 million by the end of 1994. Most—78 percent of the loan volume—went to buy land; 16 percent went for agricultural improvements such as buildings, wells, fences; and the rest bought depreciable property such as machinery or breeding stock.

Greiner points out that the loans can be made for a wide variety of agricultural enterprises. "We've done turkey and mink loans," he says. "We could do horses. We could do an elk loan."

Aggie Bonds aren't used by the nation's Farm Credit System, which raises its loan funds by selling bonds on the bond market to investors.

Privately, some critics of aggie bond programs say that the borrowers they help probably would have gotten loans anyway, and they add that lenders don't always pass on to borrowers all of the tax break they get from the tax exempt status of the bonds.

It's true that borrowers may have to be more savvy negotiators with lenders than they might have to with direct loans from the federal Farmers Home Administration. But there is little doubt that aggie bonds have helped improve the cash flow for thousands of young farmers who might otherwise find buying their first parcel of land extremely difficult.

And the programs have been well run, with very low default rates. At The First National Conference for Beginning Farmers and Ranchers, Greiner could point with pride to how the program had worked in Iowa.

"A typical applicant is a farm couple that may have farmed five years, sometimes 15 or 20," Greiner said. "Maybe these people have farmed for a period of time, have built up some good equity in livestock and machinery and use our program to buy their first piece of land. It's working very well for them." The failure rate on Iowa's aggie bond loans is a low 2.6 percent, he says. "The bank makes credit worthiness decisions. The people coming into our programs are not that highly leveraged because of the net worth requirement."

CHAPTER 9

Banks: A Good One Can Keep You From Getting Too Much Debt

Ray and Mary Ann Meismer, whose savings in their home equity helped finance a move from the city back to Ray's boyhood farm. An Illinois guarantee program enabled them to get a bank loan. Photo by Russ Munn.

Farmers have long had a love-hate relationship with bankers. In ranch country, cattle raisers often joke that if a banker wants to lend you money to buy cattle, that's probably a good time to cut back. In the 1980s, there was sometimes more hate than love from farmers who were losing their land in foreclosures. Some bankers, along with university economists and agribusinesses, had been cheerleaders to the inflationary land boom that preceded the crash in land prices during the 1980s. Since then, both borrowers and small-town lenders have become more professional, and more conservative and cautious.

Sooner or later, nearly every healthy farm or ranch that is growing enough to provide a living income will borrow from a bank or commercial lender. What do bankers look for in a young borrower? What can a bank offer to help make a new farm or ranch viable?

Jeff Plagge, president of First National Bank of Waverly, Iowa, who has long been concerned about the loss of younger farmers during the farm debt crisis, is typical of today's agricultural lenders. He wants to see more young people getting started in farming, but he knows that the income they'll earn today isn't going to pay off a highly leveraged operation—one that has financed nearly all of the resources the farmer is using. So Plagge prefers to see young farmers start out with fairly small loans.

First, he advises, borrow only for "inputs"—the actual supplies you'll need to grow a crop, for example, the machinery fuel, seed, fertilizer and other chemicals that might be needed. Later, as the farming operation generates some profits and savings, borrow for machinery or livestock. A loan to buy land is low on Plagge's list of essentials for a young farmer.

"People may be thinking they have to buy the land first. My idea of getting started is to be able to *control* assets, not *own* assets," Plagge says. "Land is, quite honestly, last on my list of something someone wants to own. So many young farmers want to dive in head first and borrow for machinery and borrow for land. Everything has to work or else you start going backward pretty fast."

Making payments on land raises the cost of farming to a very risky level that makes it difficult for a young farmer to survive a bad crop year, he says. For Plagge's advice to work, he concedes that "it's tied to the theory that you can rent land." If not, then ownership becomes more important—but no less risky.

Obviously, getting started without owning machinery, land or expensive livestock buildings means that a young farmer who doesn't get a low-interest loan from Farmers Home Administration or an aggie bond program will need help from an older farmer. That's a common situation for Plagge's customers, both at his bank in Waverly and at a previous employer, the First State Bank of Webster City, Iowa. Both banks are in northern Iowa, where good cropland is often too expensive to be in reach of many young borrowers.

Roughly 10 to 15 percent of Plagge's farm customers are borrowers younger than 35, which is about the same ratio of young farmers in the general population of full-time commercial farmers.

"I would say the typical operation we've always seen is someone coming out of their parents' operation," Plagge says. The parents may co-sign a note when their son or daughter starts farming independently. That means the parents are responsible for seeing that the loan is repaid if their child defaults. Often, parents provide other kinds of help, allowing their child to use their farm machinery on the child's rented land, in return for the young person coming back to the home farm to provide unpaid labor. The parents may help the young person rent land; they may share breeding livestock, allowing the young farmer to keep part of the offspring in return for labor or pasture. There are a host of time-honored methods that don't require a big outlay of cash but can be very helpful to the young person.

There is a second group that's getting started in a similar way, but not with parents' help, Plagge says. "They get fixed up with the right boss, is what they do. Quite honestly, in the last three years [the early 1990s] we've seen more of that than I've expected."

Older farmers without children who want to farm are hiring young people to work for them, with the understanding that they can either work into part of the business, or get some help starting their own farm. The boss may rent land from a second older farmer for the young person to farm. Or the boss may lend machinery to the younger person to farm additional land rented outside of the boss's farm.

"That is turning a little bit of a carrot for some of these guys who are trying to hire somebody, which, again, is a trend I like," Plagge says. "That's one of the realistic ways to get started without piling up a lot of debt."

In a few cases—Plagge can think of a half-dozen in farming neighborhoods he knows best—young farmers are actually working into ownership of the farm, or perhaps the livestock operation on a grain farm.

"Most [older] guys just aren't ready to give it up all at one time. If they can bring in somebody good—let's say it's 1,000 acres—maybe the young farmer starts farming 160 of that." Eventually, as the older farmer nears retirement, he might be farming only 160 acres of the farm. "One thing about farmers, they don't retire well. This is a nice way to transition out of farming."

Such arrangements aren't common yet, and they don't take place overnight, Plagge adds. "They didn't start the marriage on day one. There was a courtship." Often, the young farmer works as a laborer on the farm for three or four years, while the older and younger persons see if they work well together and if they can trust each other. There is a risk that the older farmer will make promises just to keep the hired help, or that the young farmer won't deliver on his own promises to work hard. But Plagge hasn't seen that. "The ones I've worked with have been very carefully thought through." And they are usually sealed by written buy/sell agreements or other types of contracts that specify how the farm will be turned over to the new owner or operator.

Although such business agreements are less common than parents helping their children, Plagge has been surprised at the recent interest in them. These new business relationships are exactly what Farm-On, Land Link and other organized state matching services are promoting. But the ones Plagge knows developed on a local level, without any formal intervention by such a matching service. "I think it's happening because it's being talked about so much today," he says.

With some help from an older person, a parent or a boss, a young farmer has a chance to get a bank loan for inputs—the operating costs for his or her fledgling business. Such a loan might cover the expenses of growing a crop, for example.

On a typical input loan, Plagge's bank and most agricultural lenders will finance 100 percent of the costs. The borrower makes no down payment, or sharing of the costs. But that kind of 100 percent loan is much less risky for the bank than financing 100 percent of a full-time farm or ranch. A complete farm operating loan might include a line of credit for family living expenses. And, if a young farmer is trying to borrow for machinery and to pay an installment on the cash rent of a farm, then such operating loans are "pretty difficult ... workability becomes an even bigger issue." With a simple input loan to be repaid at harvest or after the crop is sold, the bank takes less risk. A farmer might borrow to spend $150 an acre on the direct expenses of planting a corn crop, for example, but in a good year, his gross income from selling the crop might be $250 to $300 an acre, Plagge says.

If a young farmer does well and starts to build up some equity, then Plagge's bank is willing to consider loans for some pieces of machinery. Typically, the bank will require between 10 and 25 percent down on a machinery loan. And usually, the young farmer "incorporates with someone [older] and buys one or two items." In those cases, the young farmer is working in a larger family operation, or with a boss who is bringing him or her into the business.

It's rare for a young farmer to buy a whole line of machinery needed to run a farm, Plagge says. He knows of only one, and the young farmer didn't get a bank loan. He got a direct loan from Farmers Home Administration, the federal lending agency that takes on more risky loans. The young person had worked for about five years for an older farmer with 1,000 acres who was retiring. "The arrangement was, 'You buy all of my machinery and you can rent my land on crop share [for a share of the crop].' It was a pretty nice opportunity."

When a young farmer finally gets around to buying land, the minimum down payment is usually 25 percent, a requirement of both state and federal bank regulations, Plagge says. And in north central Iowa, some of the best cropland in the world, that's usually not enough down payment on land that typically costs $2,000 an acre or more. "At today's land values it won't cash flow at 25 percent down," Plagge says. A farmer often has to

put 40 percent down before the income from the crop will pay for the remaining debt at normal commercial interest rates.

There are ways though, that a bank can lower the interest rates on both
machinery and land loans so that some younger farmers with only modest
equity can get bank loans.

One is the aggie bond program, covered in the previous chapter. The
other is Farmers Home Administration guarantees of bank loans. Under
Presidents Ronald Reagan and George Bush, the federal lending agency
changed from one that works mainly with direct loans to farmers to an
agency that mainly guarantees private bank loans. In other words, if the
bank loans aren't paid back by the farmer, the federal government reimburses the bank for most of the loan—about 90 percent. The emphasis has
not changed much in the first years of the Administration of President Bill
Clinton. In fact, Clinton asked Congress for increases for some types of
loan guarantees while proposing to reduce most of the direct loan funds.
Budget-cutting pressure forced Congress to reduce spending for most of
Farmers Home Administration's loan programs for farmers. But the guarantee programs were still far bigger than the direct loans in 1995. Congress
budgeted about four times as much for Farmers Home Administration
guarantees of bank operating loans (which can finance inputs, livestock
and machinery). Farmers Home Administration's tiny $78 million alloca-
tion for direct loans to buy farmland (which includes beginning farmer
down payment loans) was only one-seventh as big as guarantees for bank
loans made to buy land. And banks have never made all of the guaranteed
loans allowed in the budget, according to Ferd Hoefner, Washington lobbyist for the Sustainable Agriculture Coalition (representing several advocacy groups, including the Center for Rural Affairs).

Nevertheless, country banks that depend on agricultural loans do use
the guarantee program. Farmers Home guarantees can be used to back input loans, Plagge says. Some banks will lower the interest rate if they have
less risk because of the guarantee, others won't. But the guarantee can
make a lender more willing to finance the farm of a young person without
a lot of backing. "You look at the people factor real hard in those kinds of
deals," Plagge says. "Is this the kind of person who can make it?" The borrower has to be willing to work hard, to sacrifice, and has to have the
knowledge and ability to keep production costs low. And, as with all loans,
it still has to "cash flow."

With a Farmers Home Administration guarantee, some banks will finance from 90 percent to the entire cost of machinery, Plagge adds. In
some cases, the loan guarantee may make a land loan possible at 25 per-

cent down. But loan guarantees still don't bring bank loans within reach of all young farmers.

"The ones that don't have anybody behind them, probably realistically, the only way they can start is with a direct loan from Farmers Home," he says. Plagge seems to prefer the older direct loans from Farmers Home Administration, which are easier to cash flow than the new beginning farmer down payment loans.

Philip Burns, president of Farmers & Merchants National Bank in West Point, Nebraska, is another advocate of using Farmers Home Administration loan guarantees. "My bank is a big user of Farmers Home Administration guarantees on loans," he said at The First National Conference for Beginning Farmers and Ranchers. "It's truly unfortunate that Farmers Home lending has a negative connotation to it. It's a government program and people have visions of huge stacks of paperwork and that sort of thing. Farmers Home has, in fact, decreased the amount of paperwork they require. ... Most banks have drastically increased paperwork. Really there's not that much disparity anymore on paperwork." Farmers Home has also eased its own cash flow requirements for guaranteed loans, making it easier for borrowers to qualify, Burns says. "For some of you people, the only way you'll ever be able to get a foothold in agriculture is through some kind of government program, so don't automatically cut it out of your minds," he told the young farmers.

Some states are beginning or considering starting their own guarantees of bank loans for young farmers. Illinois was the first, running its program out of the Illinois Farm Development Authority, the same agency that runs the aggie bond program in Illinois.

With a $10 million loss reserve budgeted by the state of Illinois, it has the authority to guarantee $35 million in young farmer loans. In a little more than half of the program's first year, beginning in April 1993, the program had guaranteed $3 million in loans made by country banks and Farm Credit Services in Illinois. The state program guarantees 85 percent of a bank's loan to a beginning farmer, who must put 20 percent down. The loans can be used for land, machinery and breeding livestock. Borrowers need net worth of at least $10,000 but no more than $250,000. The loans are limited to $300,000.

Ray and Mary Ann Meismer of Washburn, Illinois, were among the first to use the program. "Banks are looking for 50 percent equity. I can't do anything remotely close to that," Ray told *Successful Farming*. The Meismers moved to Ray's parents' 160-acre farm after living for 20 years in cities, where Ray worked for the Farm Credit System. The home farm has only 68 acres of tillable land. They were able to buy another 144 acres with better-quality land, thanks to a bank loan guaranteed by the Illinois Young Farmer Guarantee Program. Their equity from the sale of their home in Denver helped them make the 20 percent down payment.

Ray is extremely conservative. He bought a line of used four-row equipment for only $24,000 and planned to cut his costs by using no tillage. He also planned to farrow 20 sows and run a 30-cow herd on his parents' pasture land. He was renting the farm from his parents.

Still, with a modest down payment and a spartan approach to farming, the farm is leveraged enough that the Meismers expected to have to subsidize the whole operation by about $2,000 a year for the first five years. The couple and their three young children are living off Mary Ann's income as an occupational therapy manager for the northern half of Illinois for a health care provider. At the end of the five years, their annual loan payments drop from $30,000 to a more manageable $19,250.

The Meismers were among several beginning farmers who spoke at The First National Conference for Beginning Farmers and Ranchers. "We have kind of leveraged ourselves to the hilt to buy this farm," Ray conceded. "There are ways that we are stretching our capital, however. We purchased a farm that maybe was not quite the top quality farm for the area. ... The buildings were kind of run down. They weren't unusable, but they're going to take some work to put back into serviceability. Fences need to be repaired. But those are the kinds of things where we can stretch our abilities and use our strengths ... and probably get a quicker return with very little investment. I'm not afraid to buy something like that and put sweat equity into it." Ray also said he planned "to use aggressive risk management practices, trying to develop a marketing strategy where I'm going to be covering my costs on the grain that we're planning to sell, planning to use crop insurance and the farm programs to the best advantage to insure that we don't have a total disaster. We don't have much room for any disasters because of the leverage that we have taken on."

Being so leveraged that a farmer has to subsidize the purchase of farmland is nothing new. Established farmers often do that to buy a parcel of land that may be a once-in-a-lifetime opportunity. The rest of their farm generates a profit, enough to subsidize the land purchased. But the income from crops grown only on the recently purchased parcel of land may not be enough to make all of the payments on that parcel.

The fact that banks sometimes lend to older farmers to buy land that won't cash flow angered some of the young farmers who attended a workshop of commercial lenders at The First National Conference for Beginning Farmers and Ranchers. Those bank loans contribute to bidding the prices out of reach of young farmers.

Chad McDonald, vice president of Brenton State Bank of Jefferson, Iowa, drew some of that criticism when McDonald showed a detailed comparison of two farms recently purchased in his area. The first farm, with a proven corn yield of 125 bushels an acre, sold for about $1,815 an acre. The farmer put 40 percent down and financed $1,090 an acre. On the 15-year loan with a 7.25 percent interest rate, the farmer's payment for principal and interest was $156 an acre. Adding in his property taxes, his total annual payments are $175 an acre. The farmer could rent land in that county for no more than $140 an acre. "You're going to say this farm does not cash flow by itself, and you're correct," McDonald said. "If you want to look at it on a per-bushel cost, that's $1.40 per bushel on corn just to service the debt and taxes." Add in a typical cost of $1.50 a bushel for seed, fertilizer and other expenses, and the farmer pays $2.90 a bushel to grow his crop. Market prices are rarely that high, and even with government programs the economics of the purchase are risky.

But McDonald's point wasn't to brag about that loan. Instead, he was comparing it to another farm which a 33-year old producer who was already farming had purchased. The land was classified as "highly erodible" by the federal government, which means that it's subject to "conservation compliance" regulations. In short, anyone who farms that land will have to take steps to slow erosion or forego getting crop subsidies from the government. The younger farmer believed he could do that by using "no-till," a technique of planting without tilling the soil. Because it was less attractive to farmers and investors, the young producer was able to buy it for $1,160 an acre. Even though its yield was a little lower, its cash flow was better. The farmer paid $99.98 an acre on his land loan and another $17 an acre in taxes. Even with the lower yield, the debt payments cost $1.03 per bushel. That purchase looked better, from the point of view of making a living from the land. The more expensive farm's value may appreciate faster, which makes it attractive for investors. But for real farmers, the less expensive erodible land can sometimes work well, if it doesn't require more costly terraces and grassed waterways to meet the government's conservation requirements.

"It's similar to buying a Chevrolet instead of a Cadillac," McDonald said. "When you consider buying a farm, get all the details. Understand what you're buying."

Banker Jeff Plagge agrees with McDonald that more rolling, erosion-prone land can be a better way to start out—if it can be protected by the planting techniques the farmer uses and if it still produces good crops. And, he concedes that some banks are indeed making loans that, by themselves, might appear to be highly leveraged. In general, Plagge is uncomfortable with leveraging. That's one of the reasons he has mixed feelings about Farmers Home Administration's new beginning farmer down payment loans. "It's exciting to see the program come in, but on the other hand, it's a pretty leveraged deal." That's because the young farmer has to pay the 30 percent down payment loan back over a short time period of 10 years, which makes the payments higher than they would be for other direct loans from the government agency.

Young farmers may also have access to another, less regulated source of credit—loans from agricultural suppliers and farm equipment dealers. Many large agribusinesses own lending firms known as "captive finance companies." Just as car companies finance consumer purchases of automobiles, farm machinery dealers have long been willing to make loans to buy their own equipment. But in the 1980s, as banks and other traditional farm lenders became more cautious, cooperatives and other farm suppliers started doing more of the financing of inputs needed to run a farm or ranch.

"That's probably one of the biggest changes I've seen take place," says banker Plagge. "That's going to be one of the more prevalent things in the 1990s. It's more of a marketing tool than anything else."

One of those newer captive finance companies is the Cooperative Finance Association, affiliated with the large regional supply and meat processing cooperative, Farmland Industries of Kansas City. In the 1980s, the Cooperative Finance Association shifted from lending only to the local co-op members of Farmland and began lending directly to farmers and ranchers who buy supplies from those co-ops. Larry Bayer is a field loan officer for the association who works out of Norfolk, Nebraska.

The main purpose of all captive finance companies is to help their parent businesses or co-ops sell products, Bayer explained at The First National Conference for Beginning Farmers and Ranchers.

"The key is that they're not a profit center," he said. "Doing it that way, they can fluctuate the price of the product to offset the interest cost. In other words, they really have two products, the loan and the product. They can subsidize the loan or do whatever they want and still make money on the product side. So, they've really got two ways of pricing their product. ... A feed company might have a 2.9 percent interest rate. They might increase the cost of their feed by $5 a ton."

Obviously, this makes comparison shopping twice as challenging for a young farmer, who will have to make sure that both the interest rate and the price for the product will work in his or her cash flow. But both Plagge and Bayer say that beginning farmers may sometimes find it easier to get credit from captive finance companies than from banks.

"Captive finance companies are really in business just on the short-term end," Bayer says. "If we loan money on fertilizer and chemicals, we've only got to wait until the fall and, when the crop is sold, we get our money right back. ... The analogy for this is dating versus marriage. Banks marry their customers. They're in it for the long-term relationship. Captive finance companies, we only date."

"Due to this relationship, I think the captive finance companies are willing to take on more risk," he adds. "If we're selling a product, if something goes wrong, all we have to do is get our assets back and then we're back in business when we resell it. An example is a repossessed tractor. The tractor's going to look better in an equipment lot when it comes back than it would in a bank parking lot. So the equipment dealers aren't as afraid to get the product back because they can send the product right back out the door."

Bayer agrees with Plagge that captive finance companies are a significant factor in farm lending. He says that equipment companies made $4 billion in loans in 1993, enough to finance half of the new and used equipment sold. Most regional co-ops and seed and fertilizer companies also offer crop input financing.

Captive finance companies can be a useful tool for providing some credit, if the interest rates and prices are favorable, but a "dating" relationship won't be a panacea for starting-out young farmers.

As Plagge sees it, young farmers face three barriers. The first is finding land to farm or ranch in the first place, a tough job when they compete with established farmers who have more capital and who have a reputation that makes them seem less risky to landlords or landowners selling on contract. The second is finding capital to begin buying machinery or livestock to work the land. And the third is getting the farm or ranch to grow to the size that it has enough capacity to provide a decent living for a family.

In the beginning, that last requirement often can't be met by farming or ranching alone, and at least one member of the family has to have an off-farm job. Plagge says that many people complain about such a heavy demand on their time and energy.

"When I look back at my father and people in their forties, fifties and sixties, most of them started that way," he says. His father worked full time in a meatpacking plant and started out farming part time on the farm where Plagge grew up near Latimer, Iowa. Plagge's father gradually shifted to full-time farming. He knows it's not easy, but he believes it's still possible today.

CHAPTER 10

The Farm Credit System: A Farmer-owned Lender Revives Help for Beginners

Moe Russell, of Omaha Farm Credit Services, helped run one new program that the farmer-owned lending cooperative now provides for beginning farmers. Photo by Michael Malone, courtesy of *Successful Farming.*

In 1994, after seven years of working for the Mennonite Central Committee in Haiti, Marty and Heidi Gingerich and their two children, Annette and Elana, returned to Marty's southeast Iowa home, where Marty had always dreamed of farming on his own. In Haiti, Marty, 31, had been an extension worker in soil conservation and soil fertility, helping local farmers. Heidi, a registered nurse, was a rural health care worker. It was rewarding and they stayed years longer than they first intended. But as their oldest daughter, Annette, approached five, they decided that returning to the United States would be best for their children's education. It was a decision made before political turmoil and violence sent thousands of native Haitians fleeing to this country and U.S. troops to Haiti to restore civilian government.

When Marty and Heidi began farming in March, they had some advantages that came from growing up in the area, but before long they also had to find a source of credit in order to give their diversified farm enough production capacity to support a family.

95

They rented 330 acres near Wellman, Iowa, from a great uncle who also owned a cow herd. Marty planted 90 acres of corn, 20 acres of beans and 30 acres of oats on a traditional crop share arrangement, splitting the crop 50/50 with his uncle. In return for putting up hay from about 60 acres and pasturing his uncle's cattle on about 130 acres, he gets to keep one-third of the calves weaned from the cow herd.

Marty's real bread and butter is a purebred Yorkshire sow herd, which he owns himself while renting some of the buildings he needs for farrowing and concrete lots for finishing from his uncle. He has a ready-made market for the breeding gilts he raises, the Yorkshire-Hampshire swine herd of his father, Don Gingerich of Parnell, a past president of the National Pork Producers Council who is well-known in the hog industry.

Even with that kind of help, Marty soon discovered that he and Heidi would have to find a source of credit. He had originally planned to farrow two groups of 25 sows, but his cash flow projections showed that he would need a third group. Marty had taken a financial management course offered by the Iowa Pork Producers Association that's called the "Pork College." One of the financial measurements the course advocates is a "25 percent rule" that suggests allocating one-fourth of a farm's net income for living expenses and taxes, another fourth for the retiring debt, the third fourth for paying interest and the last fourth for new investment. Marty's projections showed that the two groups of sows wouldn't produce enough income to allow for growth of his farm, so he decided to add a third group, with plans to farrow about 18 sows every seven weeks.

To do that, Marty planned to buy a 12-year-old 13-foot by 32-foot pig nursery that cost $10,000. He also needed to buy a used tractor and needed additional operating money for the first year. He approached both a local bank and the local Production Credit Association office, the lending arm of the Farm Credit System that makes short-term loans to farmers who become members of the co-operatively owned lending system. Marty's mother works at the PCA office, so he knew something about Farm Credit System programs. (Farm Credit rules bar her from working on Marty's loan application and he had no special advantage.)

Marty found both the bank and the PCA willing to work with him. The local lending environment is competitive and both lenders are looking for new business. At the bank, he applied for a loan backed by tax-exempt aggie bonds. The interest rate at the time would have been between 7 and 7.5 percent. He also began the application process for a PCA loan. On May 18, he settled for the PCA loan. Rates were going up at the time, so he locked in 8 percent. Even though the PCA rate was slightly higher, his interest expense turned out to be lower. That's because the aggie bond loan would be made in a lump sum that would begin accruing interest immediately. He could draw on the PCA loan as needed and pay interest only on the money he had actually borrowed.

With a five-year loan, Marty borrowed $27,000 for the needed capital investments, the $10,000 pig nursery, $5,000 for setting up the nursery, and $12,000 for the tractor. The Gingerich couple also borrowed for operating funds for part of their first year in production—after investing about $30,000 themselves in the business. "We put some money into the operation. It was what we had," he says.

"They were real flexible in setting this up," he adds. "The operating loan money, there are parts of it that can be put into the [longer] term loan, like the breeding stock we're buying." In 1995, they planned to convert part of the operating loan into midterm notes payable over four or five years. They had been approved for a $60,000 operating loan, based on cash flow projections, and they used most of that in 1994—for operating costs, machinery, hog equipment and improvements. With sales of hogs beginning in February, Marty expected a smaller operating loan in 1995.

Marty recalls that his loan officer at first said that he would have to get special permission from the home office in Omaha, which serves a four state district of the Farm Credit System—Iowa, Nebraska, South Dakota and Wyoming. But at about the time of the loan application, the farmers elected to the board of the Omaha Farm Credit Bank had approved a new loan program for beginning farmers under 35 years of age that had slightly more lenient credit requirements.

The bank, whose trade name is Farm Credit Services of the Midlands, began its new "Beginning Farmer Program" in late March of 1994. It set aside $20 million out of its $4 billion loan portfolio for making new loans to young farmers who wouldn't meet the bank's normal lending requirements. Each loan office had $250,000 to work with and was encouraged to loan no more than $75,000 to each young customer.

The capital requirements are more lenient for the beginning farmer loans. After closing, the beginning farmer must have minimum equity of 35 percent, compared to 50 percent normally required for established farmers. Working capital has a guideline ratio of 1.3 to 1 instead of a more normal 3 to 1. (Working capital is defined as current assets minus current liabilities. Current assets are cash and things that can easily be sold for cash, such as stored grain. Current liabilities are loans and bills due within the year.)

The program's operating loans are secured with liens on crops or livestock. Longer term loans of up to five years are available for buying machinery, livestock and buildings. Those require a 20 percent minimum down payment. The program doesn't make loans to buy land, but operating loans can be made to rent land.

Although the credit requirements are more lenient than normal, the program is fairly conservative. The borrowers must have agricultural experience and background and must be currently involved in agricultural production. They have to show that their record keeping is above average and they must have crop or livestock production that is average or above.

The program is designed for farmers under 35 years of age, although the bank can't legally discriminate by age and would likely make exceptions. (At the end of 1994, the average age of farmers in the loan program was 25.5.) Borrowers can't have assumed full control of a farm for more than five years, however.

By January 31, 1995, the program had loaned a total of $1.19 million to 43 young customers. Considering that the program got a late start, after many farmers would have applied for credit, the response had been good, says Moe Russell, division president of branch lending for Farm Credit Services of the Midlands. And the local offices seemed to be following the goals of the board members.

"As I looked at the loans I was really impressed by how a number of them are beginning people with limited worth," Russell says.

The Omaha district of the nationwide Farm Credit System was apparently the first to formally set aside funds for beginning farmer loans. But, according to 1981 regulations established by the Farm Credit Administration, the federal regulator of the Farm Credit System, all of the farmer-owned banks in the lending system are to give "consideration" to "the special credit needs of young, beginning, or small farmers, ranchers, and producers or harvesters of aquatic products."

In practice, most lenders within the Farm Credit System haven't felt that they had the capital reserves to take the risks associated with beginning farmer loans until recently.

"There's some risk in beginning farmer programs. If it's not successful, everybody wants to point the finger at the lender," Russell says.

The Farm Credit System was the target of considerable finger pointing in the 1980s, when the farm debt crisis put severe financial stress on the lender, requiring a federal bailout which was later repaid ahead of schedule. Farmers considered the Farm Credit System to be less cautious about its borrowers' financial condition than private banks but more cautious than Farmers Home Administration. The system was also criticized for making risky loans to agricultural developers who sold land to absentee investors or who developed marginal land. A Production Credit Association based in Valentine, Nebraska, folded in 1984 when a single, bankrupt farmer and land developer held nearly $6 million of the Association's total loan portfolio of $70 million.

With tougher federal regulation of its loan portfolio, painful liquidation of its most troubled farm loans, and a gradual improvement in the farm economy, the Farm Credit System was nearing financial soundness by the early 1990s. Very little of the federal bailout funds were actually used and most of those had been repaid.

Unlike commercial banks, the Farm Credit System is tied only to the fortunes of agriculture, which made many of its members, especially the regional banks in the Midwest and South, vulnerable to the severe economic downturn in agriculture in the 1980s. The System can't accept savings deposits. It raises money for loans by selling bonds on the bond market. The bonds aren't guaranteed by the federal government, but because the System was established by Congress and is periodically renewed by federal legislation, it has an aura of security to investors. Income to holders of Farm Credit System bonds is also exempt from state taxation. Federal law restricts its ability to make nonfarm loans. Any business loans made by the System must go to businesses with an agricultural connection.

Congress began the Farm Credit System with the establishment in 1916 of the Federal Land Bank, which was intended to provide more competition to banks in making long-term loans to farmers and ranchers for land. The Production Credit Associations were created during the Great Depression to provide more short-term credit to farmers.

By the early 1970s, just as the boom in U.S. grain exports was starting, the Farm Credit System began a "Young Farmer Program" to encourage its associations to lend to young borrowers. By 1980, just before the farm debt crisis, Chuck Hassebrook of the Center for Rural Affairs testified before a congressional committee that the System had abandoned any extraordinary efforts to help struggling young farmers.

As the debt crisis deepened, programs for young farmers were all but forgotten, in spite of a new, 1981 regulation requiring the Farm Credit System to give special consideration to younger borrowers.

Today, the Omaha Farm Credit Services program may represent a renewed interest in encouraging the traditional borrowers of the Farm Credit System. But there are many competing forces within the System. Many of the banks from the System's original 12 regional districts have merged. The districts of Omaha and Spokane are now one lending institution, as are the three districts of St. Louis; Louisville, Kentucky; and St. Paul, Minnesota. Whether they remain responsive to their borrower members remains to be seen. Already, some Iowa members have criticized the Omaha bank for seeking to make loans to agribusinesses rather than traditional operators of farms. And in Washington, lobbyists for the Farm Credit System were seeking greater authority from Congress to make "rural development" loans that might have more to do with rural industries than farms.

The Omaha district, too, is making loans to farmers who put up hog buildings for such vertically integrated hog producing companies as Murphy Family Farms of Rose Hill, North Carolina. It doesn't make loans directly to Murphy, the nation's biggest owners of hogs, but it does lend to farmers who buy buildings for finishing hogs owned by Murphy, says Moe Russell.

"We aren't so much interested in getting in the middle of that issue," Russell says. "We want to finance agriculture as it is, not as we would like it to be."

In reality, the Farm Credit System may have little choice but to seek a wide variety of agricultural loans in order to keep a significant share of the agricultural lending market. The System was the biggest agricultural lender before the farm debt crisis of the 1980s and now ranks behind commercial banks.

But the ultimate owners of the System are still the farmers who borrow from it and become stockholders in the process. And it's possible that they will see that the System isn't likely to survive if it ignores young independent farmers. If that's the case, farmer directors of banks in other Farm Credit System districts may follow the example set by Omaha's beginning farmer program.

Marty Gingerich is well aware of the changes taking place in the hog industry and in agriculture in general. But he remains optimistic that he can survive and prosper, partly because he has chosen to specialize in raising breeding stock.

"I don't know how I would do it just raising market hogs, the size that I am," he says. "One of us would be working somewhere else."

He also thinks that the relatively small amount that the Omaha Farm Credit Bank has set aside for beginning farmers—$20 million— has the potential to help a lot of young farmers.

It will if they have the same cautious attitude about debt as Marty Gingerich. He admits that debt makes him "uneasy."

"It's nice to have money available but I think it should be used with some discretion," he says.

Traditionally, farmers have tried to buy land first to farm, he says, but that's opposite the approach taken by other businesses, he says.

"When you start another business, you spend for things that are more liquid first, then for the non-liquid assets," he says. That's one of the reasons he isn't trying to buy land right away. "I think things have to pay for themselves."

Even with that cautious approach, the Gingerich couple found the first year challenging. In 1994 they had projected that their hogs would be sold for $43. By 1995, they had reduced the projected price to $38. Heidi had taken a part-time job as a tutor for the public schools.

Even at the lower hog prices, they expect to have $10,000 to pay on the principal of their loans in 1995, after deducting for family living expenses. But "we don't have much cushion to handle any major setbacks, such as disease problems," Marty says.

CHAPTER 11

Apprenticeships: Working for Experience More Than for Money

Susan Planck (*right*) works alongside an apprentice on the Plancks' Purcellville, Virginia, produce farm. Photo by Dan Looker.

The job opportunities listed each year by the North East Workers on Organic Farms (NEWOOF) placement service in Belchertown, Massachusetts, have a slightly romanticized, New Age tone. Most of the more than 40 listings for 1994 were from New England, but they included farms in California and Malaysia. One, a four-acre spice and herb farm in Arizona, advertised "low impact living in a beautiful setting" with a "large library with much discussion on social change." Apprentices there earn room and vegetarian board and have access to "many lovely hiking trails."

What the listings don't always point out is that the apprentices sometimes aren't paid and that even when they are, their wages for 55 or 60 hours of hard field labor may be well below the federal minimum of $4.25 an hour. Yet, many young people with no experience in agriculture—usually college students or recent graduates—find these jobs appealing. On some farms, they have returned for a second or third summer of work.

In 1994 Michelle Wiggins, a 23-year-old graduate of Northeast Missouri State University who grew up in St. Louis, found herself working as an apprentice at Delta Farm near Amherst, Massachusetts. Wiggins eventually hopes to become a farmer herself.

"There's a ladder that I see that you have to climb from being a farmer wannabe to being a farmer," says Wiggins, who has a degree in biology. A step on that ladder is practical experience. Students bound for a career in law or medicine can get experience as law clerks in courts and as hospital interns. But students who didn't grow up on farms won't find a clear path to a profession in production agriculture.

"I came out to this area because I could not find any way to get training in the Midwest," Wiggins says of the New England area. Before that, she had worked one summer as an intern for Monsanto Corporation, working with scientists who were testing new herbicides in greenhouses and field plots. In 1993 she got a summer job at Ohio State University's demonstration farm near Reynoldsburg. She found the work at Monsanto interesting and enjoyed working with skilled scientists, but it only reinforced her view that chemical-based farming can't be sustained. In Ohio, she learned to drive a tractor and other practical skills. But not until she worked as an apprentice at Delta Farm did Wiggins get a chance to make the kind of production and marketing decisions that are important to surviving as an independent farmer.

The farm's owners already had established a half-acre of strawberries for the pick-your-own market, but because they were just starting in vegetable production, Wiggins had more independence than most apprentices. She did much of the planting of Delta Farm's four acres of vegetables herself in May and June. She also helped with marketing because the vegetable crop yielded so well that all of it couldn't be sold at the farm owners' roadside stand.

"We really had to scramble to find markets. That was a real job, contacting restaurants and trying to find out what to charge," she says.

With organic farming especially, learning all that you need to know from a book is impossible, she adds. "An internship is really a good way to go if you want to learn practical skills."

Even so, Wiggins sees her apprenticeship as just one more rung on the career ladder to farming. Organic farmers are able to make a living on only a few acres if they grow vegetables, fruits and flowers that can be sold in urban areas. But Wiggins knows that she'll need more money than she now has if she's going to start out renting or buying her own organic farm, even on a small scale. In the summer of 1994 she had one of the better apprenticeships offered through NEWOOF, earning $2,000 plus a small allowance for housing and food. "At some point, I'm going to have to find a job that makes money," she says.

Although Wiggins had a good experience, she knows that not all students who have tried apprenticeships have been pleased. Some discover that the farmer doesn't teach them much and begin to feel like a source of cheap labor and nothing else.

"It's real easy for an apprentice to feel exploited," she says. "Apprenticeships really are risky for the farmers and for the apprentices. I'd like to see some guidelines set up."

That opinion is shared by Judith Gillan, executive director of the New England Small Farm Institute, a nonprofit group that took over running the NEWOOF apprenticeship program in 1985. It maintains the program by charging a small fee to farmers in the listing and selling the list to some 250 potential apprentices each year.

"Many of us, and I'm among them, feel it's time the organic community moved forward and structured this more," Gillan says. Right now, Gillan's organization assumes no liability for the program and it's up to the farmers and students to negotiate working arrangements that satisfy both parties. Gillan says that the Institute is considering improving the program by offering mediation between workers and farmers, offering more formal training on employer responsibilities under the law, and perhaps forming a cooperative for farmers to buy liability insurance and workers compensation.

Most of the farms in the NEWOOF listing are small enough to be exempt from the federal Fair Labor Standards Act, which regulates wages paid to workers. The law mandates that any employer, including farmers, must pay the $4.25-an-hour minimum wage if their business used more than 500 "man days" of labor during any quarter of the previous year. That gender-biased term, *man-day*, means at least one hour of paid work a day but it could also be a full 8-hour day or more. To put it in more understandable terms, if a farmer employs more than about seven full-time workers in one quarter of the calendar year, he or she probably would have to pay minimum wage, says David Birchem of the Department of Labor's Wage and Hour Division in Des Moines, Iowa.

Larger farms can legally pay slightly less than minimum wage if they provide lodging and food. The farmer must calculate a value for the room and board and use it as a credit to bring the weekly wage up to the federal minimum. Also, farmers aren't required to pay overtime for more than 40 hours of work. Farms can get a "student certificate" allowing them to pay

85 percent of minimum wage but the students can't work more than 20 hours a week during the school year or more than 40 hours a week in the summer—requirements that would be unrealistic on most produce farms during the busy season.

Some state labor laws are now tougher than the federal laws on the minimum farm wage, says Phil Martin, an agricultural economist at the University of California-Davis. As recently as five years ago, Texas required farm laborers to be paid only $1.35 an hour. But such lenient requirements are vanishing, he says. "In the top five labor states, the tendency is to treat farm laborers as you would any other laborers," Martin says. The bottom line here, is that any farmer considering hiring apprentices would be wise to consult state and federal labor agencies first. And, if you're a student considering an apprenticeship, you're likely to feel better about the experience if you're being paid something close to the minimum wage along with your board and room.

Currently, farmers and apprentices who have used the NEWOOF program recommend at least three things to make an apprenticeship succeed:

• The prospective apprentice should visit the farm before agreeing to work there. Wiggins advises visiting several farms and comparing them. "Do they just want cheap labor or are they going to teach you about farming?"

• Both the farmer and apprentice should spell out their expectations, preferably by putting them in a written agreement.

• During the apprenticeship, both parties should talk often and frankly about whether their expectations are being met.

Some of the points to consider in an agreement between the farmer and apprentice include:

1. Hours and type of work, days off, length of stay on the farm
2. Wages, if any
3. Living arrangements, domestic chores expected
4. Level of experience of the apprentice
5. Skills or aspects of farming the apprentice wants to learn
6. Policy on use and care of farm equipment
7. Safety of the conditions and nature of work and insurance coverage
8. Sensitivity to stereotyping work by gender
9. Sensitivity to the need for privacy for both the apprentice and farmer

The NEWOOF listing of apprenticeships encourages farmers to offer a variety of learning experiences for their workers. And it reminds would-be apprentices that "farm life can be isolating" and that "hard physical work—harder than you may be used to—may be required."

One of the better opportunities on the NEWOOF list of organic apprenticeships may be Flickerville Mt. Farm and Groundhog Ranch in Warfordsburg, Pennsylvania. The farm was started in 1983 by two former Washington Post reporters, Cass Peterson and Ward Sinclair. (Sinclair died in late winter, 1995.) The couple began farming full time in 1989 and by 1990 they tried using three apprentices.

"It's allowed us to continue to farm," Sinclair says. "Rarely do we get a person skilled in truck farm requirements." So, the answer to that shortage of skilled labor was to train their own. The farm's 10 acres of vegetables, herbs and flowers and its small fruits and orchards make a labor-intensive business that grosses $100,000 in sales revenue annually. Its owners budget about $30,000 of that each year for labor costs. By 1994 the Flickerville labor force had grown to seven apprentice field hands—five men and two women, a professional chef, and a woman who began as an apprentice and now lives on the farm and works there ten months of the year, helping to manage the field crew in summer.

Many of the apprentices are college students—from Penn State or Antioch University. One is a high school student who lives on a neighboring dairy farm and is interested in seeing his family diversify their farm. And one is a former apprentice who is now renting land from another dairy farmer to raise his own produce in his spare time. With that experience, the young man may eventually be able to get a Farmers Home Administration loan to raise vegetables completely independently. "It's our way of helping him get started in truck farming," Sinclair says.

From the employer's point of view, the interests and skills of the apprentices vary a lot. "Each year we have one who thinks they want to be a farmer," Sinclair says. There are others who, "although I've never had them admit it directly, you can tell they're here on a lark—by the way they work, or don't work."

Interns are paid a flat $500 per month, plus another $100 per month if they complete their agreement to work all summer or longer. Apprentices are expected to work six 10-hour days. Sinclair and Peterson give hiring preference to those who can work more than just three months, to help with spring planting or fall harvest, or both. They deduct income tax withholding and social security. They can't afford to buy health insurance.

The farm does provide all of the in-season produce you can eat and housing, either in one of its five 8-by-8-foot sleeping cabins or in a bunkhouse that's part of a central bathhouse and dining area. Workers must do their own cooking and wash their own clothes at a laundromat in town. They can't use the farm's phone, except in an emergency. Sinclair says that the farm hires only single people since it doesn't have facilities for families with children. More expensive housing would be needed if the farm used migrant workers instead of interns. With housing and board included, the apprentices probably do make minimum wage, he says.

"In the case of having college people who are walking into this with their eyes wide open, I don't think it's exploitation," he adds.

In fact, he prefers that prospective apprentices visit the farm. And Sinclair and Peterson send each potential apprentice a four-page flier that lays out exactly what the apprentices are getting into.

They will learn a great deal about organic vegetable production. "You name it, we probably grow it," Sinclair says. "We try everything here, from globe artichokes to zucchinis and everything in between." The farm grows 50 varieties of flowers, 25 sweet pepper varieties and 20 kinds of lettuce, which the farm sells from April through November.

The farm's self-description in its flier is blunt about the demands of picking and selling such a variety of produce:

"When it's raining too much to be in the fields, your daily pace will slacken. We use that time to tend to greenhouse tasks and other housekeeping chores. But if a market is looming, the harvest must go on, no matter the weather. That sometimes means working in the rain, cold, intense heat, semi-darkness. It is neither easy nor enjoyable."

There isn't time for formal training seminars. Instead, "we gather everyone around and show them how to use a hoe or how to transplant a lettuce seedling or whatever we're doing," Sinclair says. Not everyone runs machinery. "Some people will take a tiller out and not put oil in it or even check the oil," he says. He and Peterson do nearly all of the planting of seeds, because if they don't germinate, they've lost too much time. They work alongside their apprentices and put in longer hours than they do.

"It's a different crew for us every year and it's a different experience for us and a different challenge," he says. "We've made a number of friends. We keep hearing from people who worked here."

"Having young people around keeps you current," he adds. "I've learned a lot of new words and to wear my ball cap backwards."

Sinclair has also learned the hard way that it doesn't work to hire "pre-formed couples." They once lost two apprentices during a busy time because the couple had broken up. Their recruiting flier also points out:

"We do not run a police state, but as a matter of policy, because of unhappy past experience, we expect you to honor our no-drugs-on-the-farm rule."

Sinclair concedes that his own view of labor relations has changed a bit from his newspaper days, when he says he was a liberal. "As a business person, I'm sounding more and more like Jesse Helms," he says, referring to the U.S. senator from North Carolina. "We jump through so many regulatory hoops that seem unnecessary—and it tends to make you conservative."

Yet, Sinclair doesn't think it's right for organic farms to hire apprentices without any compensation. Some organic farms that are making little money try to attract people to work there without wages, he says, "and it's kind of an exploitive situation."

When looking through the listings of farms accepting apprentices, look first at "a farm that pays real money for work," he says. Some red flags are listings that say "lodging is available in the community," or "stipend negotiable" or that say the apprentices will live with the family or that say "some baby sitting included," he says.

Another farm that pays reasonably well and attracts students to work more than one summer is Wheatland Vegetable Farms of Purcellville, Virginia. The farm, owned by Chip and Susan Planck, is not listed in the NEWOOF directory. Instead, the Planck's keep a detailed description of their farm, work requirements, pay and accommodations on file at the placement offices of a half-dozen colleges in the eastern half of the country.

Chip and Susan raise 45 acres of vegetables, including heirloom varieties such as the Cherokee Purple tomato. They sell in 15 farmers markets in the Washington, D.C., area. The farm is similar to organic farms but uses a low-toxicity insecticide on some crops at the seedling stage only. The Plancks describe it as fitting the "low-input, sustainable agriculture" category, with pesticides (mostly biological) and fertilizers making up only 2 percent of their costs.

They employ up to 14 workers at any given time during the growing season, Susan says. "I'd say half of our people are probably second or third year and are making above minimum wage," she says. "The first-year people are not." First-year students start at $3.50 per hour and get a 25 cent-an-hour raise after two or three weeks. The Plancks provide free but rustic housing, in small rooms that have been partitioned out of the back side of an old hay barn. Workers also get free produce from the farm and generally spend about $15 a week on food. Second-year employees start at $4.50 per hour and third-year workers start at $5.50.

Apprentices store tomatoes that will soon be taken from the Plancks' farm to farmers markets in the Washington, D.C., area. Photo by Dan Looker.

The Plancks themselves earn roughly $50,000 a year, but both work seven days a week in the growing season and do all of the hard physical work of their employees. From the years 1989 through 1993, their hourly pay was one to three times that of their lowest-paid workers, a more egalitarian ratio than that of even such progressive companies as Ben and Jerry's, the well-publicized ice cream company.

Students average about 55 hours of work each week, Susan says. "It's a lot of dollars a week with virtually no expenses. Everybody goes away completely surprised by how big a check they have."

Jean Mountford, one of the Plancks' employees in 1994, agreed with Susan that her net earnings at the end of the summer turned out to be much better than she might earn in a city, with the transportation and housing costs that would go along with urban life. On the farm, "you don't pay any rent and you pay like $5 a week for food, so you end up saving so much money."

But beyond the pay, it was the valuable experience that kept Mountford, a student at Wesleyan University, coming back to Wheatland Farms. In 1994 she was working there her third year, just after graduating with a degree in math and religion. The Plancks give each worker responsibility for loading vegetables into trucks and taking them to one of the farmers markets where Wheatland Vegetable Farms sells. The students then sell the produce on their own.

"We're in charge of our own markets, which is so great," she says. Mountford and two other students plan to start their own vegetable farm on 15 acres of land that another student's father was going to let them use on his upstate New York dairy farm. Wheatland Vegetable Farms was a good place to learn her new occupation, Mountford says. "It's so great to have someone who trusts you and gives you responsibility."

Anyone interested in apprenticeships listed by the NEWOOF Apprenticeship Placement Service should write or telephone:

NEWOOF/New England Small Farm Institute
P.O. Box 608
Belchertown, MA 01007
413/323-4531

Young people not interested in organic agriculture do have more opportunities to learn about conventional agriculture through traditional education, either at colleges of agriculture in land-grant universities or at two-year community colleges that offer an agricultural curriculum. Both the two-year and four-year schools have placement offices where, as we will see in the next section of this book, farmers sometimes join large agribusiness companies in recruiting employees.

A few private recruiting firms also place high school, community college and university graduates in agricultural jobs, on large commercial dairies, big feedlots and with large swine companies. The firms also have a smaller number of openings for jobs on diversified crop and livestock farms. Usually, the recruiting firms collect a fee from the employer, based on between 15 and 25 percent of the employee's first-year salary. If the job applicants don't have a degree in agriculture, the recruiting firms require a farm background.

Agra Placements, Ltd. of West Des Moines, Iowa, specializes in recruiting for agribusiness companies that sell feed, seed, chemicals, fertilizer and farm machinery, says recruiter Gary Follmer. The firm also has some job listings for swine companies, horticulture, and food processing.

Rarely does it have on-farm jobs listed, he says. In agribusiness, it even normally has a few international listings, including ones in France and in Mexico. The firm was founded in 1974 and has four offices:

2200 N. Kickapoo, Suite 2
Lincoln, IL 62656
217/735-4373

4949 Pleasant, Suite 1
West Des Moines, IA 50266
515/225-6562

16 E. Fifth Street
Berkshire Court
Peru, IN 46970
317/472-1988

Valley Office Park
10800 Lyndale Ave. S., Suite 214
Minneapolis, MN 55420
612/881-3692

A competitor, which Follmer credits with offering more on-farm jobs, is AGRIcareers, Inc., which bills itself as the oldest agricultural recruiting firm, founded in 1968. The company has two offices, in two small Iowa towns:

P.O. Box 140
Massena, IA 50853
712/779-3744

or New Hampton, IA 50659
515/394-3145

AGRIcareers has job listings on dairy farms, swine operations, feed-lots, large cow-calf ranches and farms with livestock and grains, says Nancy Erickson, farm production specialist at the firm's Massena, Iowa, office. In 1994, opportunities were limited in dairy and swine production while more listings were offered in beef production and on livestock-grain farms, she said.

A qualified candidate, with farm experience or training, typically earns $18,000 to $20,000 a year as a full-time "farm assistant." (Recruiters don't use the term, "hired hand.") Compensation varies, depending on whether the farm employer furnishes housing.

At any one time, a qualified candidate might have about a dozen potential jobs to choose from, she says. Some farmer-employers will list the chance to work into their farm or participate in a buyout, she says. Others may have that in mind, but won't list it. Farmers who want to offer young employees a chance to buy their business or farm in partnership, usually wait until the employee has been on the farm between two and five years, Erickson says.

But many of the job seekers do have that goal in mind, she adds. "It's everybody's dream to be self-employed and to be farming on their own."

Farmers
Helping
Themselves

CHAPTER 12

Recruiting: How One Visionary Farm Family Sought a Successor

Dwight and Sally Puttmann recruited Joe Hlas as a college student to work with them on their Kingsley, Iowa, farm. Joe's wife Julie has joined the Puttmann's second family. Photo by George Ceolla.

Like many farm couples nearing retirement, Dwight and Sally Puttmann of Kingsley, Iowa, want to preserve a lifetime of stewardship. "I guess we put together a farming operation we're proud of," says Dwight, a soft-spoken methodical man who doesn't brag much. He and Sally hope that whoever follows them will keep the terraces and grassed waterways on the hills. They don't want the livestock buildings idle.

The Puttmanns, who have farmed since 1957, currently raise 700 acres of corn, soybeans, and oats and run 30 beef cows on 40 acres of pasture. They sell 2,500 market hogs annually. Dwight is also one of seven owners of a 13,000-pig-per-year feeder pig co-op.

With two daughters married and living in cities, the easiest choice at retirement would be to auction off their machinery and rent their land. "I could have whispered over the fence that I wanted to cash rent it, and all I'd have to do is wait until the first of March every year for payment," Dwight says.

That doesn't appeal to Dwight or to Sally, who has served on Iowa's soil conservation committee. She doesn't like the way some renters "farm up and down the hill." The Puttmanns have been active in Farm Bureau and their church, and their view of how the land should be treated has been influenced by their religion. "We believe that the land is just yours to take care of," says Sally, a cheerful woman with an infectious laugh.

Selling the land wasn't a pleasant option, either. Their life's work with cattle and hogs might not be continued, and their two daughters still wanted the land to stay in the family. So the Puttmanns decided to find someone who would eventually be able to manage the farm.

They could have helped one of many nieces and nephews but, "if things didn't work out, there'd be a split" in the family, Dwight adds. "If we didn't get along and it blew up in our faces, it would be a family feud for the rest of our lives." So they decided to bring in a stranger.

Several years passed before they acted on that decision. They didn't think their farm was large enough at the time to support two families. So they gradually bought more land and they increased the size of their hog business. "That took a while to get all of that in place, even before we started the interviewing process" in the winter of 1987, Sally recalls.

Recruiting a stranger to eventually work into their business isn't a new idea. In fact, they got the idea from a friend. But in 1987, before any of the formal matching programs between unrelated older and young farm families were started, the Puttmanns seemed visionary. They were also altruistic. They decided to help start out a young person with little capital instead of working in partnership with an established operator.

Today, farmers with a similar point of view can find potential successors through established matching programs. But even now the Puttmanns' approach is worth studying for those families who don't live in a state with a matching program—or for those who just prefer a do-it-yourself approach.

The Puttmanns considered three ways of finding someone. They could have advertised in the local paper. They could have gone through an agricultural employment agency, which would charge them a fee. Or they could work through the agricultural placement service at a college.

They chose the ag placement office at Iowa State University because they thought that candidates there already would be studying a discipline that could help them succeed in farming. They also picked ISU because Dwight, who taught vocational agriculture early in his career, is an alumnus. They worked with one other school in addition to ISU—a community college at Iowa Falls, which has a reputation in the state for having a good training program for work in the swine industry.

For the placement office at ISU, the Puttmanns filled in a form with a brief job description for someone with the ability to help raise crops and who liked working with livestock. They were looking for a college junior who wanted a "summer job with possible long-term opportunity." The listing was posted at the placement offices of the two schools in January.

Thirteen ISU students turned in resumes, which the placement office sent to the Puttmanns. They got another resume from Iowa Falls. They eliminated three or four who seemed to have other opportunities to farm and narrowed the list to the top seven, the number they thought they could interview in a day.

In early March, they drove to Ames, Iowa, and conducted the interviews at ISU, devoting an hour to each one. They also met the Iowa Falls student at an Ames restaurant.

During each interview, "we told the student he would be working with us and not for us and that we wouldn't ask him to do anything we would not do," Sally says.

They began the interviews by introducing themselves and describing their farm and their family. They gave background about themselves and their interests. And they brought pictures that showed the enterprises on their farm.

Then they asked each student to describe his career plans. They explained why they were holding the interview and their own long-range plans for the farm.

They asked each candidate to explain why he was applying for the job and to rate himself on his ability to work with hogs, to maintain machinery, to weld, in carpentry and to put electrical wiring in buildings. They also asked about the candidate's interest in community activities.

They learned later, that some of their questions aren't allowed under the federal Equal Employment Opportunity Act. Most of the federal law applies only to businesses with 15 or more employees, but state laws may apply, so it would be wise for farmers to check on this with their state civil rights commission or agency that administers equal employment laws. Some of the questions the Puttmanns asked that could not be asked by large employers included whether the student had a history of substance abuse. They had asked if the candidate was married, because that would make a difference in the kind of on-farm housing they would need to provide. And they asked about his church affiliation, believing that would show more about the candidate's character.

Finally, they asked if the student had any questions of them and they ended the interview by promising to notify the candidate of their decision in two weeks.

"At the end of the day, Dwight told me to drive home. He said he was going to sit in the car and just think about the interviews," Sally recalls.

It wasn't an easy decision because several candidates appeared strong. In the end, they settled on Joe Hlas of Tama, Iowa, partly because they thought he had a strong desire to farm. They also learned that his mother's farm wasn't big enough to support both Joe and his older brother. Joe's father died in 1983 of brain cancer and his older brother had taken over the family's 550-acre rented farm near Tama, Iowa. Joe's good grades at the university also showed that he was an excellent student.

Joe spent the next summer helping Dwight with field work and hog chores. It was a trial period to see how well the two could work together. At summer's end, the Puttmanns offered Joe a permanent job after college and gave him a few months to think it over.

"I told them right then and there, 'I'll be back,'" Joe recalls. "They were nice people. They were easy to work for. You never found Dwight in the coffee shop. He was always working."

Today, Joe farms 230 acres that he rents on his own, using Dwight's machinery. The Puttmanns pay him a salary to help Dwight on their farm. They also pay Joe a bonus of 2 percent of their gross receipts to help him build up equity. The farm's receipts vary from $250,000 to $500,000 a year, so that bonus can be as much as $10,000. Someday, Joe may rent the farm when the Puttmanns' daughters inherit it.

To others contemplating bringing in a young farmer from outside the community, Sally cautions that "you may encounter a little resentment—both against the young person and the people bringing him in. Number one, you'll probably have relatives who are envious. I know there was a little resentment for a short time."

But Joe managed to help unruffle some feathers. "Dwight served nine years on the local school board. When you do that, you make a few enemies," Sally adds. "One woman who had never spoken to me since then was charmed by Joe later on. Many people in our church were glad to see Joe, too. So things do have a way of turning around."

When Joe first moved to the farm, the Puttmanns provided a trailer that "was quite big compared to the dorm room I lived in," Joe says. He helped Dwight build a new home during the winter of 1988-89. Joe moved into the old farmstead after that and was joined by his bride, Julie, when they were married in 1992. When he was single, the Puttmanns provided Joe noon meals.

"It used to amaze me when I was dating Joe how well this guy was taken care of," says Joe's wife, Julie, who recalls that Joe's refrigerator was well stocked with Sally's leftovers.

The Puttmann family even played a role in Joe's marriage. Julie, who worked for a time in public relations for the Minneapolis Police Department, met Joe when she was visiting her parents in LeMars, Iowa, one Thanksgiving. One of the Puttmanns' daughters, who lives in LeMars, set Joe up with Julie on a blind date.

Besides gardening and helping Joe, Julie runs her own computer desktop publishing business, Personal Publications, out of their home.

Joe has also used the computer to help modernize the Puttmanns' record-keeping system. Sally points out that sharing financial information was an important way to establish open communications when Joe moved to the farm.

"One of the first things we did, we turned the shoe box of records over to Joe—and that's literally what we did. We felt it was important for him to understand how much income we had and to understand the farm's finances," Sally says. After Joe computerized the records, "we also felt that we needed to keep a handle on what was going on, so we have continued our own hand-written records."

Joe says he could have made a higher salary after graduating from college. His on-farm pay and bonuses normally are $15,000 to $20,000 a year. But he considers the Puttmanns extremely generous. Dwight pays Joe wages to work the 80 acres he rents from Dwight on a crop-share basis. It seems like double dipping, since most renters wouldn't be paid a salary. And each year, the Puttmanns move Joe closer to independence. The first year, when an attorney offered to rent 80 acres to Dwight, Dwight rented it to Joe.

The next year, Dwight bought another 80 acres that he rented to Joe. The third year, he helped remodel a barn on Joe's farmstead for farrowing. The fourth year, he rented 90 acres to Joe that Dwight had rented from his brother. The fifth year, Joe had saved enough to buy a tractor and 60 acres of the 80 owned by the attorney. Joe farrows 25-to-30 sows a year and sells feeder pigs to Dwight. Dwight loans Joe the sows; Joe supplies the feed and labor, then sells pigs to Dwight to finish. Dwight also takes back the sows and sells them.

At Dwight's side, Joe has learned farrowing, carpentry and an electrician's skills. And, most important, he says, he's learned time management.

"I'll be thinking, 'We need to get this done,' and he's already decided we'll do this next Friday," Joe says. "College is nice background, but it is not even close to what it's like working on the farm."

Dwight says the arrangement would be a sacrifice for those who want to winter in Miami and summer in Minnesota. But the Puttmanns think they'd be bored with that. Instead, they've added to their family. And Joe and Julie feel very much a part of the family. Since moving to the farm, Sally has helped Julie, taking care not to be too intrusive. Julie appreciates Sally's gardening tips, for example. "Sally's as much of a mother figure to me as Dwight is a father figure to Joe," Julie says.

"Believe me, the two of us, we don't feel like we're giving up anything," Sally says. Dwight adds that "what we have done here is not the cheapest and is not the easiest. If you don't enjoy seeing a young person progress, it may not be for you."

Dwight clearly does enjoy putting his old teaching skills to the ultimate practical test. Joe recalls with some amazement that on his first summer at the farm, Dwight "hardly knew me, yet he gave me a lot of trust." Joe grew up on a relatively flat farm in east central Iowa, and when he first cultivated corn on the contour on the Puttmann's more rolling farm, he was nervous about it. Dwight took Joe to the field and left him on a tractor to start cultivating. "Eight rows were laying over after a short stretch," Joe recalls, "but Dwight was driving away."

To those farmers who like what the Puttmanns have done, the couple says it's important to plan early, preferably between age 50 and 55. Dwight was 61 when Joe was in his fifth year on the farm. If you wait until you're 65, "you don't have the energy to think, 'I'm going to spend five years training someone,'" he says.

Passing a farm to the next generation takes patience and hard work on all sides. But Dwight Puttmann believes there can be more matches with beginning farmers.

"I'd like to see a lot of it being done," the Iowa farmer says. "I sit here and see people retiring and there are not many people moving onto these farms. There are people gobbling up another quarter-section. You wonder where it's all going to end."

Others could do this, adds Joe, but "maybe it's one chance in a million the way it's worked out here."

Even as generous as the Puttmanns are, they have been cautious about writing a formal agreement about exactly how the control of the farm will be transferred to Joe when they retire.

"We have discussed our long-range plans with Joe," Dwight says, "but I am reluctant to put anything in writing. I have another farm that I may rent to Joe in a year or two, which would bring the amount of land Joe farms on his own up to about 400 acres. Eventually, Joe plans to buy my machinery. If I get out of farming, I plan to sell my shares in the feeder pig co-op to Joe."

Adds Sally, "In four years we'll be at retirement age. We will be look-ing at a buyout agreement with Joe for our equipment. It will be on a con-tract of some kind, to string out the expense for Joe and to help the tax sit-uation for ourselves."

Ultimately, the Puttmanns may change their will so that when their daughters inherit the land, Joe would have the right of first refusal if the daughters decide to sell the farm. That would give Joe some assurance that he would have a chance to continue using the land if he could no longer rent it. They have no formal agreement that if the daughters don't sell the land, Joe could continue to rent it. But Sally doubts that Joe would lose that opportunity as long as her daughters own the land. "They get along so well. I know they feel really good about him farming with us," she says.

Dwight has been impressed by Joe's ability to progress on the farm and in his own independent farming operation.

"When I first started with Joe, I asked, 'Do you want to go fast or go slow?' He said 'What do you mean?' I said 'Well, Joe, if you want to go fast and take over the machinery and the livestock, it'll be about $300,000. I can bury you good, and you'll be there forever. Or, if you want to go slow, I can take you into this fairly debt free," Dwight recalls. "Joe was more conservative than I am, so he wanted to go the slow route and Joe has been basically debt free until he got into the land purchasing and machinery pur-chasing. He started from nothing and is farming 240 acres for the last three years and never borrowed a cent. That says something for him."

Sadly, Dwight Puttmann died in September 1995; Joe Hlas is currently sharing more farm management responsibilities with Sally Puttmann.

CHAPTER 13

Sharing Opportunity: How One Farmer Helps Many

Bob Dickey (shown speaking to *Successful Farming*'s second Beginning Farmer Conference) has helped several young people get started in farming. Photo by Ed Lallo, courtesy of *Successful Farming*.

Bob Dickey of Laurel, Nebraska, has been successful in his farming career. He and his wife, Mary, run a farm that grows 500 acres of corn and 500 acres of soybeans, making it large for the area. And his 900-sow feeder-pig business, although not huge in the rapidly changing hog industry, is still quite a bit larger than average. He pastures 100 stock cows. He's also active in his church and has served on both the Nebraska Corn Board and the U.S. Feed Grains Council.

Conventional thinking suggests that young farmers would have a hard time benefiting from larger farms. Most young people don't have the resources to buy or rent them. And in many rural neighborhoods, large farms are seen as the competition, often bidding against young people who try to rent a small parcel of land. Some economists would argue that farms like those of Bob Dickey will eventually become part of even bigger businesses.

Bob Dickey defies that conventional wisdom. Bob and Mary have three children in high school or college and the farm may stay in family hands. So he's not working a younger person into his business as a potential owner. Instead this northeast Nebraska farmer spreads pieces of his business around among several young people. In the fall of 1994, he was contracting with three young men to finish out some of his feeder pigs for sale as market hogs. And he was helping another young man who rents 160 acres of irrigated farmland from Dickey to also build up a cow herd. Finally, Dickey lets an employee use his farm machinery on land that the employee is buying for his own farming business.

Dickey doesn't view this as a sacrifice, even though he conceded that in the fall of 1994, when market hogs in his area were bringing as little as 32 cents a pound, he could have made more money selling all of his hogs as feeder pigs instead of contracting to have young people finish them out. Dickey sells some of his feeder pigs through an electronic auction run by Farmland Industries, the regional farmer-owned cooperative based in Kansas City, Missouri.

"As far as I'm concerned, we're going to hang in there," he said then. He expected prices to eventually improve as other pork producers are discouraged by low income and leave the business. And he was preparing for the future by switching to genetic stock that will produce hogs that yield 2 to 4 percent more lean meat than average.

Both the older and the younger farmer must benefit for any of these business arrangements to succeed, he believes. As he explained at The First National Conference for Beginning Farmers and Ranchers in Omaha in March, 1994, "In my opinion, there's one thing that all leases, all buy/sell agreements and all other entry arrangements must have in common to be successful and that is a win/win situation. The Land Link program and other similar programs are very worthwhile, in that it's a win/win for young people, it's a win/win for middle-aged farmers such as myself, and it's a win/win for those farmers who want to retire."

For middle-aged farmers like Dickey, business arrangements that help young people get started on their own are advantageous because "it allows people like me to hire and retain competent employees." The best, most capable young people with agricultural skills often have a goal of farming on their own. For the young people, there can be advantages to working with an older farmer who has more experience. And for retiring farmers, there are tax savings to gradually shifting ownership of their business to another generation.

Dickey has worked with young people from a variety of backgrounds. "Some have jobs in town and want to supplement their city income with additional farm income," he says. "Some are just beginning farming and don't have a sizable operation to be fully employed and they're looking for some way to supplement their income."

"In my case, I'm looking for this type of young person who lives on a farm, who's not fully employed and has hog facilities. I don't have enough room to finish all the pigs that we raise from our 900 sows, so I contract with these young farmers in the area to feed out my pigs," he explains. "I pay all the feed expenses, the vet bills and so on, and the young farmer provides the labor and facilities. One of the young farmers I contract with has 450 pigs. I pay him seven cents per head per day to take care of these pigs. If he doesn't lose any pigs during the month, his check at the end of the month comes to $945, which is an excellent way to supplement his income."

Contract production of hogs has been criticized by some established farmers, mainly because large companies ask the farmers to invest thousands of dollars in new hog buildings, which are very efficient. The large company sells the hogs and takes the marketing risk. But if the hog company doesn't fill the buildings with hogs quickly after each batch is marketed, or if the farmer has a severe disease problem, the investment in the building can quickly become a liability. And, perhaps the biggest objection is that when prices in the market are high, contract production often is less profitable for the farmer doing the work than raising hogs on his own would be.

But in Dickey's case, he contracts with young people who use existing buildings, so there's less financial risk. And, in late 1994, with Dickey assuming the marketing risk, it looked like a particularly good deal. "That's the advantage to that young, beginning farmer," Dickey said. "He's not going to get hurt under that condition [of poor market prices]."

Dickey adds that "we use the all-in, all-out concept when we take pigs to these young contract feeders. For example, we take 450 head to this young farmer in a short period of time, say two or three weeks. At the end of four months these hogs are ready to go to market. They're all shipped over a period of two, three or four weeks, and when that last hog is sold, that's when we figure out our profit and loss statement. If it's favorable, then we bring in another 450 head for this young person." (If the markets aren't good, Dickey doesn't blame that on the young person. He's looking at how efficiently the young farmer was able to finish out the hogs and at how many pigs were lost to disease.)

Dickey's business relationship with a cattle-raising tenant is another example of how he helps young people. "I have 100 head of stock cows on contract with a young farmer in a neighboring county and he takes care of these cows for a percentage of the calf crop," he says. "This young farmer has excess labor and pasture and also I rent him an irrigated farm, on which he runs these cows over the cornstalks after he harvests. When it comes time to split the calves, I take my share home and feed them out. He can either keep his share out or he can settle at that point."

"One other thing that we've worked out, he's trying to build his own stock cow herd. He has two options with the calves. He can either sell them or finish them out," he says. " He has elected in the past, and I've allowed him to do this, to select some of the better heifer calves that are in that calf crop and he keeps those. He's building his own cow herd for the future. And I think, of course, that that's a win/win situation."

"Another example of an entry arrangement in farming is one of my own employees," Dickey adds. " This employee is buying a farm on a contract from a retired farmer that lives in town. The employee uses my machinery. He works full time for me as a salaried employee. And he manages his own farm. He's responsible for everything that goes into his farm. He has to be responsible for the soil testing. He's responsible for buying seed, fertilizer, chemicals or whatever he needs for the crop. He's also responsible for financing the inputs, which he does through a local bank."

That's typical of just about any independent farm. But Dickey has thrown in another benefit besides lending his machinery. "When the crop is harvested in the fall, I allow him to put it into my bins and whenever he wants to sell his crop, he's allowed to do that," Dickey explains. "This arrangement has allowed this employee to increase his skills in production and increase his skills in financing. He's also been able to maintain and grow in some marketing skills. I certainly don't have all the answers, but if he comes to me and asks what I think about a particular situation, I'm glad to share any experiences or expertise that I may have."

"This has been a win/win situation," Dickey says. "He's building equity in a farm that he's buying by using my machinery. And, of course, I've been able to have a good employee through this period of time. I hope that this relationship will last for many years."

Dickey's last example of a "win/win situation" is a farm that he sold 17 years earlier. "I sold a young beginning farmer, a person from Iowa, a 40-acre tract of land with a building site. There was more land on this farm, but he just wanted to start small and he wanted the building site to have as a home base. This individual worked at Iowa Beef [IBP] at the time. I sold him this farm on a contract and he paid me monthly, which he preferred. I asked him why and he said that was when he got paid from Iowa Beef. That was fine with me because then I'd know almost immediately if he was going to fall behind on his payments if he paid me monthly. I'm happy to say that person always was on time with his payments. His check was always in the mailbox on the first of the month. I'm also delighted to say that as of two years ago, I received my last payment from him. He owns that farm free and clear today."

"By the way," he adds, "he quit working for IBP ten years ago when he was able to rent some other land in that area. And he's also had enough cows and sows and sheep so that he's fully employed in his own operation."

The young farmers working with Dickey today seem pleased with their business arrangements. Ryan Lubberstedt, a Dixon, Nebraska, farmer in his mid-30s, has finished three groups of pigs for Dickey, an arrangement that seemed especially favorable in the fall of 1994 when hogs were selling for well below the cost of raising and feeding them.

"There ain't no risk involved. I got caught in a hog situation like this seven or eight years ago. I filled up with 500 pigs and paid too much and got burnt," he recalls his money-losing venture in buying and finishing feeder pigs. Lubberstedt also grows crops on 700 acres of mostly rented land that includes 115 acres he's buying from his father. He has been farming since 1978. But it's the hog finishing that's "paying all my monthly bills," he says.

He finishes hogs in a remodeled farrowing house that he bought a decade earlier for $2,400, and in a horse barn where his father installed a cement floor, and in three hog huts with cement lots. "It ain't nothing fancy, but it works," he says.

After losing money finishing hogs on his own, Lubberstedt worked with another farmer who paid him for finishing the farmer's hogs. But the farmer didn't always have enough of his own feeder pigs to fill those buildings and would buy other feeder pigs. Getting pigs from several sources dramatically increases the risk that the pigs will get sick. Lubberstedt's hogs did. "All I was doing was carrying around a syringe and a bucket," he recalls.

Dickey's hogs all come from his own farm. Lubberstedt has been pleased with the quality and his pay. "It sure helps when you've got a good clean bunch of pigs," he says.

Earl Johnson is in his late 20s. In partnership with his brother he is buying 60 acres of land that his grandfather once farmed. His working arrangement as an employee on Bob Dickey's farm makes it possible. Dickey allows Johnson to use his equipment to plant and cultivate the parcel of land the brothers are buying. Johnson's brother, who works for another farmer 40 miles away, has purchased a small, used John Deere 55 combine that harvests the crops.

The brothers remember their father losing his farm in 1968. They've taken a conservative approach to buying land. They signed a contract for deed that has a short, seven-year amortization. The brothers put up one year's payment as a down payment. They won't buy any more land until that parcel is paid off, Johnson says. "I don't plan on stopping at 60 acres, but you've got to start someplace."

Buying that land would have been difficult without being able to use Dickey's machinery. "If we had to buy all of our equipment, that would make too much overhead and too much of a risk," he says. "We're getting prices for corn and beans that my dad talked about getting back in the 60s."

Johnson and his brother had already signed the contract to buy the land when he started working for Dickey in 1989. He made use of machinery a condition of his employment. "When I interviewed, I wanted to make sure I had access to equipment wherever I went." Dickey also provides Johnson housing.

In the county to the west of Laurel, Dickey rents an irrigated 160 acres near Bloomfield, Nebraska, to Brad Doerr. Doerr, who started farming in 1979, is buying 320 acres of land on his own and rents 400 acres of pasture from a different landowner. On that pasture, Dickey and Doerr have worked out a mutually beneficial arrangement. Dickey doesn't have enough pasture for his 100 head of cows—Angus and "Black Baldy" (the local nickname for white-faced Hereford-Angus crossbred cattle).

So Doerr cares for the cattle on his rented pasture land, and Dickey pays him with a share of the calves born from the cows. For Doerr, it was a chance to diversify his farm. "I wanted to build up my cow herd but I didn't have the funds to do it," he recalls. When he started raising cattle, cows were selling for $800 to $900. Borrowing to buy a herd that size would have cost more than $1 million in principal and interest.

Doerr's farming business is now large enough that his wife no longer works off the farm and is able to spend more time caring for the couple's three children.

"I really appreciate what Bob has been doing," he says. "It would be nice to have more people like him."

In the long run, Dickey doesn't view his efforts as a sacrifice and he encourages other older farmers to consider more business ventures with young farmers. But he conceded at the Omaha conference that finding an open door to farming will require hard work and preparation by young people. He even encouraged his younger listeners to join a Toastmasters Club or to take a Dale Carnegie course "to learn to communicate better and to sell yourself. ... There's a lot of competition out there, looking for that win/win situation in agriculture."

CHAPTER 14

Working into Ownership:
A Partnership Passes Down a Dairy Farm

Doris and Ray Weber (*above*) gradually transferred their dairy herd to their employees through a partnership agreement. Debbie and Ted Mausser (*below*) found their hard work rewarded after nine years as employees. Photos by Bob Coyle.

Ted and Debbie Mausser had worked on the dairy farm of Ray and Doris Weber near Dubuque, Iowa, for almost a decade when Ray's doctor forced them into putting their plans on paper. Almost since the beginning, Ray had said that he wanted to bring the Maussers into the business somehow. The Webers have no children of their own and they were pleased with the way Ted and Debbie had worked in helping with the milking chores on the dairy farm in northeast Iowa.

But Ray hadn't stopped in his busy career to plan exactly how the transfer would take place—until 1990. Then, when he was 57, his doctor ordered him to retire. Ray had gone through his eighth knee surgery and had other health problems. That meant that Ray would receive social security disability payments, which complicated how the two families would set up their partnership. Under a simple partnership, the income from the farm would be split in half between the two families. But, in order to keep from losing his Social Security payments, Ray could not have an active role in the management or be required to put any labor into the operation.

"It took us about four months of hard study and talking to people to get things figured out," recalls Debbie, who keeps the books for the dairy farm. The good-natured mother of four jokes that "I had a headache for four months straight."

"Different people would tell you different things that might be contradictory," adds her husband, Ted, who is the manager and operator for the partnership that finally emerged from their efforts—RWTM Dairy Farm. The two couples sought advice from an attorney, a bookkeeper, an Extension Service farm management specialist for their area, and another specialist with Iowa State University who helped them run a computer projection of three years of cash flow for the partnership.

Their partnership was officially started in January 1991. By the fall of 1994, it seemed to be working well for both families, after making a few adjustments.

Debbie opened a thin file folder on the Maussers' dining room table and showed the six, relatively simple agreements that make the partnership work and that are transferring much of the farm's assets to the Maussers over ten years. Here's a brief description of each agreement:

1. A two-page "Partnership Agreement" that says that Ray, Doris, Ted and Debbie are four partners in the business. Income from the partnership is to be divided equally between the four partners. As we'll see shortly, each family is deriving income from the partnership, but the partnership itself so far has not produced income. The partnership's purpose is listed simply as "ownership and improvement of dairy herd." It has few specific duties, except to provide health insurance for each member. It can be changed at any time by mutual agreement of the members.

2. A "Dairy Herd and Farmland Management Agreement," describes Ted's responsibilities to the partnership. He's responsible for the management, labor and operation of the farm. And he can hire any additional labor that's needed. In 1994, his wife did the bookkeeping and the couple, along with help from their junior-high-age daughters Kristen and Melinda, milked about 75 cows. This is how the Mausser family derives income from the partnership. The original plan was for the partnership to pay Ted a salary of $1,700 a month. But an accountant has advised them that a partner can't be an employee of his own partnership, so Ted receives a "guaranteed payment" annually instead. At first, this agreement also required the partnership to take out worker's compensation insurance. But the partners later decided that the $1,600-a-year cost was too high. The partnership is

not required to have it and the coverage was redundant since the business has health and liability insurance.

3. Under a "Cash Farm Lease" the partnership rents Ray and Doris's 140-acre farm for $95-an-acre per year. The lease started at about $150-an-acre but Ray agreed to lower it after a couple of years of poor milk prices. Although the land rent is an important part of the income that Ray and Doris derive from the farm, Ray says he was willing to lower it to make the partnership work. As long as each family gets adequate income through the partnership, it doesn't matter how each piece of the arrangement works, he says. The partnership also rents another 143 acres from a different landowner. Ray and Doris had rented that land in their own name before turning that lease over to the partnership.

4. There is a ten-year "Equipment Lease" that the partnership uses to rent Ray's farm machinery for $12,200 per year. This is the other main source of income to the Webers from the partnership. The agreement lists all of the machinery on the farm and assigns a market value to it. It's used machinery and no one item is extremely expensive. Ray is depreciating the machinery for his own personal income tax purposes. At the end of the ten years, the Maussers will be able to buy it for a low, depreciated value. In essence, the lease transfers ownership of the machinery to the Maussers.

5. A "Bill of Sale" records the sale of half of the dairy herd to Ted and Debbie. "They worked here for nine years and, just as kind of a bonus, I turned over $104,000 of dairy cows, heifers and calves to them," Ray explains. Although the sale was essentially a gift, Ray doesn't view it that way. "When they started working here, their salary was $12,000 a year, which is not very high. That was actually in consideration of good services given to me by them." And, he adds, "If they had to buy that $104,000 worth of cows to start with, it would never have worked." After the "sale" of the cattle to the Maussers, they put their half of the herd into the partnership. Ray and Doris put their own half into the partnership, but draw no income personally from the cow's production, or the sale of culled cows or calves. Their main sources of income are the land and equipment leases to the partnership. The Maussers did not have to pay income tax on the cattle, but the donations could have counted against the $600,000 exemption from federal estate and gift taxes. Ray and Doris planned the gift carefully so that the tax effects were small. Ray and Doris each can give $10,000 apiece to Ted and Debbie annually tax-free. So they each gave $10,000 worth of cattle to Ted and to Debbie—or $40,000—at the end of 1990. Then, in early 1992, they gave them another $40,000 worth of cattle. That

gift of $80,000 over two calendar years leaves only $24,000 that counts against the estate tax exemption.

For Ted and Debbie, the gift of cattle can't be used personally as collateral for a loan because it's now in the partnership. Nor can Ted and Debbie depreciate the existing cattle on their personal income tax return.

6. Finally, another bill of sale records the sale of $50,000 worth of feed to the partnership by Ray and Doris.

None of this would work if the two families didn't already have a good relationship and trust each other. In fact, when the families drew up the partnership agreement, they used the same attorney, something the attorney was reluctant to do because the legal interests of Ray and Doris differ from those of the Maussers. "We ironed things out between ourselves quite a bit before we ever went into the lawyer," says Debbie.

There are some important documents still missing from Debbie's folder of agreements. The most important is the issue of how this transfer will be finished.

Most important, perhaps, is control of the land. A common approach to turning over control would be to sell the land to the Maussers on a contract, with Ray and Doris financing the sale. Instead, the Webers plan to will the farm to the Maussers. If Ray dies first, the land may go into a trust for the benefit of his wife. "When we die, they'll inherit everything that's left," Ray says. While the Webers are alive, they will need the rental income from the farm. That rent is expected to go up, in fact, when the machinery lease ends. "The income of $630 a month from Social Security doesn't quite feed you," Ray explains.

At the end of the ten years, it is likely that the Maussers will buy out the other half of the partnership— the cattle and the machinery. Instead of relying only on a will to transfer the land, the Webers may instead start to transfer ownership by selling the land to the Maussers on a contract. That would spread out the Webers' tax burden over the life of the contract.

Ray concedes that it would have been simpler and less worrisome to just sell the farm when he had to retire. But it would also have been a more expensive approach to getting income from the farm he had spent his lifetime building.

"I guess we took a look at what our tax obligation was going to be," he recalls. "If we sold all of this, it looked like we were going to owe our government between $200,000 and $250,000" in capital gains and income taxes.

Ray, who has served as president of his county's Dairy Herd Improvement Association and, before that, as president of the local National Farmers Organization, is proud of his herd of Red Holstein cattle. "Our nice little red and white dairy herd out here, I spent 20 to 25 years building up a registered herd of dairy cattle," he says. He wanted to see that work carried on. His final reason for helping the Maussers is that "we did not have any kids of our own—and we can't take it with us."

The machinery lease also shows the flexibility in this arrangement. The partnership has purchased some newer machinery, including a new total-mixed-ration mixer. Ray traded in an older tractor and a manure spreader and gave the partnership the full trade-in value. That's higher than the depreciated value and is more favorable for the partnership than standard leasing practices.

One thing the partnership did not do is use a buy/sell agreement. This is an agreement that gives the younger person, in a family or in a partnership like RWTM Dairy Farm, the right to buy the farm when the owner dies. It's usually written into the will. Those formal agreements are often protected by life insurance policies on the potential seller and sometimes on the buyer. A life insurance policy on the landowner with the younger farmer as the beneficiary gives the young person the capital to buy the farm, or perhaps enough for a down payment. In some cases where the younger person already owns enough livestock and machinery that it would be difficult for the older person to buy it back if he or she dies, a buy/sell agreement may also be financed by an insurance policy on the young farmer with the older farmer as the beneficiary.

Ray went as far as having his insurance agent draw up a policy, but he discovered when he saw it that "in five years it was going to cost over $10,000 a year to insure me," because of his poor health. They decided the partnership couldn't afford a buy/sell agreement that depended on an insurance policy. And, because the Webers plan to give the farm's remaining assets to the Maussers when they die, a buy/sell agreement may not be needed.

Although the families have worked well together to make the partnership viable, it isn't without any problems. At the end of the ten years, when the leased machinery is fully depreciated, the Maussers should be able to buy the Webers' half of the machinery in the partnership for a nominal amount. But buying the newer machinery like the mixer will be more costly, as will buying the other half of the dairy cows.

Both Ray and Ted expressed some concern about the fact that the partnership hasn't generated enough income to replace as much of the aging machinery as they'd like to. Ray says that some chores on the farm may have to be hired out. They already have their corn harvested by a custom combine operator. Ray thinks they may eventually have to put up silage the same way. "Our self-propelled silage chopper is pushing 25 years old," he says. "When you look at the price of a new chopper, you're looking at roughly $80,000. When you fill only three silos a year, you can't possibly do that."

And, as is often the case for young farm families, the Maussers, who are in their mid-30s, find it challenging to raise four children only on the income from the dairy farm. Debbie works part time at a flower shop to supplement their income. And she's earning a two-year degree in accounting at a nearby community college.

Both came from the same northeast Iowa rural community. And they both have farm backgrounds. But they can't count on their families for much financial help. "I'm the oldest of nine kids and I have five brothers, so dad's farm is spoken for," Debbie explains. Ted's parents don't farm, he adds, but "I spent a lot of time on my grandparents' farms. My dad left the farm when I was six or seven. I guess that I always wanted to be able to farm ..."

Debbie interrupts, smiling, "but he didn't tell me until after we were married."

Their first three years of marriage were spent in Hawaii, where Ted finished his service in the Marine Corps. After that they returned to northeast Iowa, working for a few months in a toy factory in Dyersville. Both had been laid off when Ray Weber ruptured a disc in his spine and started calling around to find an employee. The Webers had known Debbie's parents for years. Both couples belonged to the National Farmers Organization. Ted was recommended.

"Ray called me one night and I said I'd come out and help out," Ted recalls, "and it's been every day since then."

"It's just one of those dream situations for the Maussers," says Phil Hufferd, an Extension farm management specialist who first suggested using a partnership to transfer the farm between the two couples. Hufferd didn't draw up the final agreement. That was done by the lawyer that both families hired. But Hufferd is familiar with the arrangement and says it's probably one of the most successful in the state.

There is no one way to transfer a working farm from one generation to another. The Maussers and Webers could have formed a corporation instead of a partnership. But they're likely to dissolve their partnership if the Maussers are able to buy out the other half from the Webers at the end of ten years. Ending a corporation can be more expensive than ending a partnership.

There are a host of estate tax and income tax issues that need to be considered in any type of legal structure aimed at transferring a farm from one generation to the next. In many ways, the tax laws are more favorable for unrelated people, Hufferd says.

"There's a lot of things you can do being not related that you can't do if you're related," he says. For example, when the Webers nearly halved the rent they charged the partnership for their land to keep the business viable, the Internal Revenue Service likely would not allow that between a father and son without some tax consequences, Hufferd says. The rental rate would have to be close to rents charged others in the area.

By bringing someone onto the farm who is unrelated, a farm couple approaching retirement can enhance their retirement income compared to what they could expect through the traditional approach of liquidating the farm, he says. "But at the same time it's very important that you bring in people that are going to be good stewards of that investment. From the senior parties' point of view, that's their retirement income."

From the younger person's point of view, written agreements help clear up an uncertainty about how they may be rewarded for their long-term sacrifice. Agricultural employees are often paid less and have fewer benefits than they might find at a job in town, so an assurance that they're indeed working into ownership of all or a share of a farm can be a vital morale booster.

Everyone involved in RWTM Dairy Farm has a vested interest in making the business work. Ray Weber no longer helps with milking. His knees won't allow that anymore. And the partnership couldn't require him to do any work at all, or he would lose his Social Security disability payments. But he still operates some farm machinery that doesn't require "a lot of clutch work," he says—chopping silage, cultivating corn. Debbie adds: "He does a lot of volunteer work."

The Webers and Maussers may not be family, but they do look out for each other.

CHAPTER 15

Land Seekers: How One Couple's Persistence Paid Off

Steve Hopkins and Sara Andreasen at their first rented dairy farm. Renting land and leasing cows was a key to starting out with little capital. Photo by Dan Looker, courtesy of *Successful Farming*.

It wasn't an easy move. With a second child on the way and with about 20 cows to milk, Steve Hopkins and Sara Andreasen uprooted their two-year-old dairy farm in October 1994, and moved it nearly 200 miles to Newton, Iowa.

But this was all part of a long-range plan for the couple in their 30s who did not grow up on farms, and who spent roughly five years after college preparing and planning for a new career. In 1992 they got a toehold in farming by renting cows, 20 acres of pasture, and farm buildings near Decorah, Iowa, for a little more than two years. Now they were buying a 30-acre farmstead and pasture land near Newton, a town better known for its Maytag appliance factories than dairying. In fact, dairies are scarce around Newton and more common in the scenic, rolling northeast Iowa farmland around Decorah. But Steve and Sara had found a farmstead that the neighbors didn't want at a price they could afford. Now they would have more independence to practice seasonal dairying with management-intensive grazing techniques.

139

Steve and Sara's story shows the importance for young people who want to farm to have a carefully thought-out long-range plan. Neither Steve nor Sara grew up on a farm and they got started completely on their own. But in order to do that, they spent years saving a small nest egg and learning about the type of agriculture they wanted to practice—in this case, grassland farming that uses techniques refined in New Zealand. Then they systematically approached finding land, first to rent and later to buy. But, perhaps more than anything else, they have been persistent and haven't been easily discouraged. If they had been, it's doubtful that they would be farming today. The farm they bought in 1994 was just the latest example of how their persistence made farming possible.

They decided to start looking for a farm to buy in 1993. The rented farm had been a good place to start out, and its recently remodeled house was a clean, comfortable place to live. But, still, renting left the couple less control over their situation. There's no guarantee of being able to rent beyond each year. Costs aren't certain. And, they didn't own their own home, "which is kind of the American dream, and we had moved there from a situation where we had owned our own home," Steve says. When their first child, Anders Lee, was born that year, "we decided we wanted more permanence." By then, they had a production track record, "which we could use as a basis in seeking financing. And we had also seen interest rates bottom out and starting to rise, so we thought, 'now is a good time to buy.' And we felt that our money could just as easily be going into a mortgage as opposed to rent."

So they moved up their long-range plan to buy land after three to five years of renting. "When we started, our plan was to get some experience in milking and get an idea of whether it was something we wanted to do, to get a couple years behind us," Steve says. They also had more assets than when they began renting. After about six months of milking cows they began purchasing the Jersey cows they had been renting from a nearby farmer.

At first, Sara had felt more comfortable with renting, too. If they decided that farming wasn't for them, they could more easily sell livestock and equipment and move. "We decided we'd rather own animals and the assets for farming than land," said Sara, who helped Steve with milking in addition to working part time as a school social worker in the Decorah area. "If this doesn't work out, we'll move into something else. We don't feel trapped. We feel lucky."

By 1994, they had decided that they liked farming and life on a farm and were ready to make the commitment to move more permanently onto the land. At first, they began looking for farms near Decorah, where Steve and Sara had gone to undergraduate school at Luther College, and where they had many friends.

Even before deciding to buy a farm, Steve recalls, "I spent quite a bit of time just driving around the countryside, looking at farms that were owned by retired farmers who had put their land in the CRP"—the federal Conservation Reserve Program that paid landowners rent for ten years if they agreed to quit raising crops and plant the land to grass or trees. With most of the land in the CRP planted to grass, "I anticipated that those farms would be perfect for grazing," Steve explains. The ten-year contracts would begin to expire in 1995, and Steve expected some of that land to come on the market.

After deciding to look for land in 1994, the couple searched for a small farm with milking facilities for about $100,000. They hoped to buy between 40 and 80 acres. They soon discovered that nothing was priced low enough to buy 80 acres at that price. And the smaller farmsteads were expensive in the Decorah area, too. "We were basically looking for kind of a run-down place that we could fix up—and could afford to fix up," Steve says.

"It put us in competition with people wanting to own an acreage rather than a place to farm," he says. "Any decent county seat town has a strong acreage demand."

They also looked at a few farms being sold privately by retiring farmers. Most of those were 100 to 150 acres in size and a little too expensive.

Steve and Sara looked at about ten farms in the Decorah area over a year. They decided they would have to expand the area of their search. So they began looking at farms near Grinnell in central Iowa. They had both lived and worked there for about four years—Sara in social work and Steve as a county Extension specialist in sustainable agriculture. They had friends there, too. And, as in Decorah, they knew some of the local bankers, a crucial contact if they were going to get a mortgage for a farm. "Character is an issue in getting loans," Steve says.

In the spring of 1994, Steve noticed that a 37-acre farm near Newton, Iowa, that he had seen advertised in *The Des Moines Register* about eight months earlier was back on the market again. Its price had been cut by about a third, from $150,000 to their $100,000 price target. In March, on the way back to Decorah from a trip to Nebraska, Steve and Sara drove by the farm. "I didn't like it and Sara didn't like it at all," he recalls. "We had an image of buying a place that was a little more bucolic." The 15-year-old ranch house and a 90-foot metal dairy building looked too new to be picturesque, yet they were already rundown and the place was overrun with weeds.

In April, they visited another farm near Grinnell and decided to stop with the real estate broker at the farm outside Newton, anyway. "It was a cold, gray day in April, which didn't make it look any more attractive at all," he says. When they went inside the house, the rooms were dirty and unkempt and the house smelled musty. The carpet reeked of pet urine. As they drove away, Sara said, "I can't believe we stopped here."

But in that visit, Steve had seen great potential. The dairy building had a concrete milking parlor, a two-by-four herringbone design that would allow him to milk eight cows at a time. That system would eventually let him milk twice as many cows as the 20-stall tie-stall barn at his rented farmstead near Decorah. And a milking parlor, which has a low standing area for the herdsman, puts a cow's udder at eyeball level—greatly easing the back strain of milking in a barn. At winter workshops, Steve had heard a Wisconsin dairy farmer extol the advantages of a milking parlor, which that farmer was building on his own for about $10,000. Normally, a milking parlor capable of handling eight cows would cost about $80,000 new. At the Newton farm, Steve and Sara could get a whole farmstead for a little more than that price. Steve was hooked.

On the drive back to Decorah, they ruled out the other farm they had visited near Grinnell. It had no outbuildings at all and the house needed work. That left only the Newton farm. It took a few weeks to reconcile their perspectives on it, differences that may be as old as farming. As Steve concedes, "despite feminism and women's liberation and everything else, women still look at the house and the men still look at the barn."

Before making an offer, though, there was more to do. Steve had to check on the rumors and gossip that made the farm unappealing to local buyers. The seller had let his cows die of neglect and all kinds of rumors had floated through the local farm community about the place. That was one of the reasons why the price was low. "Anybody within a five mile radius knows the owner didn't take care of things," Steve says. So he checked with the previous owner's veterinarian, who assured Steve that the farm didn't have a disease or environmental problem. Its reputation was caused by the previous owner's apparent personal problems.

At about the same time, Steve called a banker he knew from his days of working in the Grinnell area and explained his financial plans. Steve said that they thought they could come up with a 10 percent down payment and he learned what type of loans the bank could make on the property. "We pretty much knew what our monthly payments would be before making a bid on the place," he recalls.

About two weeks after visiting the farm, they called the real estate broker with an offer of less than their $100,000 target. The broker called back a half-hour later to say the seller had accepted. "The reason the price dropped so much was that it fell between the cracks. It fell between the acreage market and the farm market."

There were last-minute problems. After the land was appraised, Steve and Sara learned in mid-June that the farm had only 30 acres, not 37, although the seller didn't know it. "We couldn't turn back at that point," Steve says, because Sara had quit her job near Decorah and had accepted a job in Newton. They had some difficulty getting the seller to renegotiate the price, but he finally lowered it slightly. Also, the seller, apparently to pay debts, had stripped the milking parlor of railing, milk lines and the bulk tank, so Steve had to search for good used equipment. That was no surprise, but "I had to educate myself very quickly on what sort of equipment is needed in a parlor," Steve recalls. "I spent most of my summer on the phone talking to people about equipment." He found a bulk milk tank advertised by two north central Iowa brothers who were switching from dairying to hog farming. A dairy co-op field representative told him where he could buy used bars for the parlor. After moving in, they discovered that an abandoned well that was too close to the barn to meet dairy codes had to be filled. They also had to make improvements to the existing well that they didn't expect. But all of these improvements were close to the $15,000 price range that Steve expected after checking with an Extension dairy specialist. And, as Sara puts it, "we're putting in what we wanted and we know it will work. Our goal was to get up to 40 cows, and we think we can do that here."

Financing wasn't simple, either, although they had no trouble getting a loan and were approved by two local banks. Their first loan was for part of the down payment. They planned to put 10 percent down, but had only part of that at the beginning of the summer. They knew that their herd's summer milk production would generate enough income to pay the rest of the down payment. Steve applied for an operating loan secured by his summer production and had no problem getting it from a Decorah bank whose loan officer had visited Steve and Sara's farm. No Decorah banks would make the land loan with only 10 percent down, however. All had a minimum of 25 percent down.

The couple did find two banks in the Newton area that would approve a 10 percent down payment, if it was guaranteed by Farmers Home Administration. They also had to show that they had off-farm income. The loan has 30-year terms, although Steve and Sara believe they'll be able to pay it off sooner. Their cash flow will improve after their first year on the farm, when they'll have their cows paid for. Of the two banks that approved their loan—one in Newton and one based in Kellogg, Iowa—they chose the Kellogg bank because they had known the loan officer since they had lived in Grinnell, and because the bank is independent and is likely to have a long-term commitment to making agricultural loans.

The same bank also made a seven-year short-term loan to finance the improvements to the milking parlor. It was backed by aggie bonds from the Iowa Agricultural Development Authority. "The Jasper County Farmers Home office was very supportive. They were good to work with and everything clicked," Steve recalls. Finally, they had to get a "bridge loan" from the bank to take the place of the Farmers Home Administration guarantee because the federal government wouldn't have funds for the guarantee until its new fiscal year began on October 1. As it turned out, Steve and Sara were disqualified on a technicality for a state beginning farmers loan. So they made the milkhouse improvements with a conventional loan from the Jasper-Poweshiek Security Bank.

Long-term, even though they depend on off-farm income to make the deal work, they believe that when the farm hits 40 cows, Sara may be able to work away from home less. "We think we can pay for it and not be real dependent on outside income because I don't want to have to work full time indefinitely," Sara says.

Steve believes they were able to get the farm purchase loan without much difficulty because they had good records of their first two years of production and because they had a realistic plan for the new farm.

They showed bankers the financial analysis of their first two years and a projected cash flow for the next three years, when Steve plans to increase the herd size by five cows per year. He intends to buy five bred heifers each spring. Unlike many dairy farms that retain heifers born into the herd, Steve will sell off the heifers when they're weaned and buy bred heifers later. That way, "every grazable inch of the farm is going to be grazed by a milking cow," and not by heifers not yet producing.

The financial analysis of their first two years of farming included the cost of milk production per hundredweight of milk, the cost of production per cow, net income per cow and net income for the herd. "Both banks thought our cash flow was excellent," Steve says.

Although state banking regulations in Iowa require that banks lend no more than 75 percent of the value of a farm, banks can require as little as 5 or 10 percent down if the loan is guaranteed by Farmers Home Administration, explains the banker for Steve and Sara, Tom Rude of Security Bank of Kellogg-Sully. "Most loans will not cash flow with 5 or 10 percent down," he says. But the loan for Steve and Sara did, thanks to off-farm in-come and a good cash flow from the dairy operation. Eventually, after Steve and Sara are able to build the herd up to 40 cows, his projected cash flow shows an annual net income between about $20,000 and $25,000.

In a way, the records for their first two years looked good, not only because Steve's low-cost methods are profitable, but because Steve and Sara had decided to start farming with no debt. They rented nearly every-thing they needed to start dairying—the cows, land, milking barn, even the tractor. Their biggest startup cost was $1,500 of electric fence. They also had to buy a used manure spreader. Even in their first year of dairying on the rented farm, they didn't subsidize the business. "We wanted the farm to be self-supporting," Steve recalls. And it was. During the first summer, their gross margin was more than $1,000 a month. In winter, with higher costs of buying hay, it dropped to about $500 a month. The gross margin is what's left from the checks they get for their milk after deducting feed, milking costs, veterinary bills, and rent for pasture and the cows.

Starting Out by Renting

Steve and Sara's frugal approach to getting started, and their persis-tence in finding a farm to rent, show that, while it's not easy to start out in farming without help from parents, it's not impossible.

They had begun looking for a farm to rent in 1989. It would be a three-year odyssey through Iowa, Missouri and Wisconsin that would take them twice to the farm near Decorah that they finally rented in the summer of 1992.

Steve had little formal training in agriculture. He has a master's de-gree in land resources from the University of Wisconsin, where Sara got her master's in social work. But even when Steve was dating Sara at Luther College in Decorah in 1984, he knew that he wanted to farm.

"Sara thought that was a phase I was going through," he recalls. It wasn't. Steve worked a short time on a dairy farm near the college and later at a dairy farm in Wisconsin. He spent years learning more about con-trolled grazing. When he worked for the Extension Service, he had a chance to visit New Zealand, where the technique was perfected. After Steve and Sara married in 1989, most summer vacations were devoted to learning more about controlled grazing—on trips to farms in Ohio, Penn-sylvania and Missouri.

Beyond his strong interest in dairying, that type of farming made eco-nomic sense.

"One of the things we decided that we wanted to do was get into dairy farming because, at least for the time being, we realized we would need monthly income," Sara says. "And dairy farming is one of the few things where you do have monthly income in farming."

"One of the things that Steve and I did is select geographic areas in the Midwest that we wanted to live in. Steve started contacting dairy field representatives—the guys who work for the different dairies, who pick up milk from all the farmers. And we found out about who those field people were by contacting the dairies. Steve contacted the field men to find out who were the retiring dairy farmers in the area, or who were the older farmers that were wanting to get out," Sara recalls.

In 1990, Steve and Sara learned from a dairy fieldman of a small dairy farm near Decorah that was up for rent. They applied, but the landowner turned them down. Even though Steve had some experience working part time with dairy cattle, he had little experience growing crops. The landowner thought that Steve's lack of a general farm background would be a detriment.

In December 1990, they visited the farm of Linda Terry near Greenfield, Iowa. Linda's young husband, Dixon Terry, was a well-known farm activist who had been killed by lightning. Steve and Sara had read in a newspaper article that Linda wanted her land farmed organically. Steve thought that he could meet that requirement by using controlled grazing. That method of farming relies heavily on pasture for the dairy herd's feed. Rotating the cattle from one intensively-grazed pasture to another, using electric fencing, usually controls weeds without the need for herbicides or large amounts of fertilizer.

The couple decided against renting the Terry farm when Sara found few opportunities for an off-farm job in the area.

Later, friends from a Wisconsin dairy farm where Steve had worked invited them to go into partnership, but the deal fell through when their friends' parents wouldn't go along.

In July 1991, a woman from the West Coast whose family owned land near Grinnell contacted Steve and Sara about renting their farm. That deal fell through by February 1992, after months of negotiations. She wanted the couple to sell organic milk, and didn't seem to believe Steve when he said there was no local market for it. He knew he could run a dairy farm to produce organic milk, but it would wind up being sold to a conventional dairy processor. They couldn't agree on the amount of the rent, either. Finally, they decided the woman was too demanding and eccentric. "She wanted us to plant wildflowers in the ditches," Sara recalls.

So, it was welcome news when they heard that the landlord of the farm they had visited near Decorah was again looking for a tenant. By then, the farmer had put his cropland into the federal government's Conservation Reserve Program and all that was left was a farmstead plus about 20 acres of pasture. Steve thought that was perfect for his plans to use controlled grazing. That technique has the potential to double grass production compared to the normal practice of turning cows out into a larger pasture and supplementing their feed with hay.

"This allowed us to rent a farm that no one wanted. We were finding a niche to get in. There's not much competition for a place like this," Steve explains. "For prime farmland, there is."

The landlord remained skeptical when approached for the second time. He thought Steve and Sara's budget for feeding expenses was too low.

"We had the numbers there," Steve recalls. Their prospective landlord "was being very tough and businesslike. I think he was trying to scare us, but we didn't blink."

Instead, they approached the landlord as if they were going through a job interview, with some skillful salesmanship thrown in.

"We had sent him resumes and we sent him information on grazing management," Steve says. "I had actually sent him a videotape on dairy grazing techniques to show that this technique could work and that we could do it successfully."

"Another way that allowed us to get in was that this particular landlord was used to a 50/50 type of rental arrangement, where the milk check was split in half," Steve adds. "Instead, we proposed simply going on a cash rent basis, where he was guaranteed a certain amount of income every month, regardless of how much milk we produced. That gave us the freedom to manage our dairy herd as we saw fit. It gave him the security of a guaranteed amount of income every month. It put the risk on us to make sure that we could come up with that cash."

By May, the landlord agreed to offer them a rental contract—the pasture for $25 an acre annually, plus $100 a month for the buildings and milking equipment and $350 a month for the house. In June, they moved to the farm and began leasing 16 Jersey cows, for $30 a month per cow, from another farmer in the area. The farmer who owned the cows kept the cows at his farm for a month while Steve and Sara got used to milking them. "I called it farm school, because I had never milked a cow before," Sara says.

The years at the rented farm went well. Sara worked as a social worker there, too. And Steve brought in extra income by working part time for the local Soil Conservation Service and by getting a research grant to hold field days to demonstrate controlled grazing on his farm.

Just how hard the two were willing to work was apparent on a visit to their rented farm in October of their first year at Decorah. A sharp wind swept the rolling pasture where Steve, clad in a sweatshirt and worn jeans, was unreeling electric fencing. Sara was finishing up the milking chores in the barn. After each milking Steve closed in a new section of the pasture before the couple let the cows out of the barn. As the days shortened the grass growth was slowing, so he had to plan the size of each temporary paddock carefully to provide enough grazing. It would be almost dark before Steve and Sara turned the cows into this overnight pasture. Then Steve would have to spend the evening on the phone to get his broken four-wheeler fixed. As winter approached, he would soon need it to haul more hay to the cows.

Yet, Steve and Sara seemed undaunted. "In terms of getting started in farming, almost all beginning farmers are low in money and low in capital," Steve says later. "What we have is energy and labor. We can provide labor. That's one of the reasons we chose to get into dairy farming is because it requires work. And we're willing to do that."

CHAPTER 16

Buy/Sell Agreements: Ways to Transfer Farm Ownership

Jim Ulring, a financial planner based in Decorah, Iowa, has degrees in agriculture, dairy science and agricultural economics from the University of Wisconsin. He has worked as a cattle buyer, auction company manager, appraiser and dairy herdsman. Photo courtesy of Jim Ulring.

J im Ulring of Decorah, Iowa, runs a financial planning firm, Ulring-Sekt Associates, Inc. in Decorah and Sioux City, Iowa, that helps small businesses continue operating when the owners retire or leave the organization. Ulring, who owns farms and has a degree in agricultural economics, works with farms as well as businesses. Most small businesses have plans to allow a new generation ways to take over the business, he says. Most farms don't.

In recent years, as a large number of established farmers begin planning for retirement and the settlement of their estates, the term, *buy/sell agreement* has become a buzzword.

"I don't like the term buy/sell that well," Ulring says. "I like *business continuation agreement* because that describes what it is."

"If you've got anyone involved in your business, any key person—it doesn't have to be a partner—and if you want that business to continue, you should have a business continuation agreement in place," he says.

The agreement is a legal contract that describes how the transfer will take place, Ulring says. A good agreement should cover all of the circumstances that would take the current owners out of the business—death, disability, retirement, divorce. Farming usually ranks as the occupation with the highest rates of disability, yet many of the agreements Ulring has seen don't deal with how the business will continue if the owner becomes disabled.

The most commonly used business continuation agreements fall into three categories:

1. *A stock redemption agreement.* This is an agreement between a corporation and an individual who doesn't control the company but has an interest in it through the ownership of shares. When the owners with controlling interest in the stock retire, die or leave the company, the corporation buys that person's stock, Ulring says. In effect, the person who still has some shares in the company becomes the owner. It's really the new owner who is buying out the old, but it's done through the corporation.

2. *A buyout agreement or third-party agreement.* This is an agreement that passes the farm from an individual owner to a key person (who might be an employee or a child working on the farm).

3. *A cross-purchase agreement for a partnership or a limited liability company that has several owners.* This allows the owners to buy out the interest of the owner who retires or leaves the farming business.

In all of these examples, the agreement spells out exactly how the next generation to run the business will acquire it. It gives the younger members of the business the right to buy it. It defines the mechanics of the sale. It can also set the price of the property in advance. For the owner, "it provides a guaranteed market for your business," Ulring says.

A business continuation agreement isn't worth anything without a way to pay for the business's assets—the land, machinery, livestock, grain stored on a farm.

"You build the agreement, then you have to figure out how to make the agreement work. Some young person that has $5 to his name, that's not going to work," Ulring says.

One way to fund an agreement is for the owner or the owner's farm to buy a life insurance policy, with the young farmer who will take over the business named as the beneficiary. As we saw in Chapter 14, the Webers didn't use insurance to help fund the transfer of the dairy farm to the Maussers because it was too expensive. The right type of insurance, purchased early in life, might have been feasible.

Most commonly, the insurance policy is on an older farmer with a child or key employee as the beneficiary, but in some cases, insurance may be needed on a younger member of a partnership or farming operation, says Phil Hufferd, an Iowa State University Extension management specialist who has helped farmers with business continuation agreements.

"Let's say it's a 300-cow dairy herd and the son owns half of the herd and half of the machinery. If something happens to the son and the father can't buy the other half, it may put the father in financial difficulty," Hufferd says. In that case, the family would need an insurance policy on the son with the father named as beneficiary. If the son has dependents, separate insurance to benefit them would also be needed.

The insurance proceeds can be nontaxable at death, if the agreement is structured correctly, Hufferd says, but that's not automatic. Buying insurance policies that would pay off the entire value of a farm may well be too expensive and probably isn't necessary, both Ulring and Hufferd say.

"My concern is having one that gives people an adequate down payment so they can keep the business viable without undue hardship," Hufferd says.

Ulring says a policy only needs to provide enough for a down payment of about 40-to-50 percent of the value of the business, with the rest of the purchase financed by some type of credit. "Most businesses with 50 percent down will probably survive," he says.

Although life insurance has a place in financial planning, "generally it's not the only answer" to finding a way to help the younger farmer come up with the money to acquire the business, says Ulring. Because of tax laws and the nature of the livestock business, farming offers many opportunities to fund a business continuation agreement that benefits both the older and younger members of the business, he says.

For example, let's say that an older farmer buys a new tractor. It would be too expensive for a younger farmer without a lot of capital to buy. So, the older farmer depreciates the tractor over the seven years allowed by the Internal Revenue Service, reducing his income taxes in the process. At the end of those seven years, the tractor is worth nothing for tax purposes, "but I can guarantee you that thing is worth $45,000 or $50,000, and you can move that over to the younger farmer," Ulring says.

Of course, because the tractor still has value in the marketplace, the older farmer could sell it to someone else—but he'd have to pay income tax on the proceeds from the sale. Let's say the farmer gets $30,000 for the tractor, Ulring says. His state and federal income taxes might take 35 percent of that, leaving him with a net of $19,500 after paying $10,500 in taxes. Instead, the farmer could simply sell the tractor to a son or key employee for the $19,500 at retirement (or a maybe a little higher to cover some of the income taxes he'll still have to pay on the $19,500).

Livestock works very well as a funding mechanism that builds a younger farmer's equity and reduces the tax consequences for the older farmer who phases out of ownership. With a cow herd, for example, a young farmer can be paid with the offspring, or a share of the calves. When the older cows are no longer productive, the older farmer sells them as culled cows, not the more valuable breeding stock that would be sold at a herd liquidation sale upon the farmer's retirement.

Another example would be a lease of a dairy herd, Hufferd says. The older farmer might continue to own the cows and give the calves to the young farmer. Over time, the owner depreciates the cows down to a salvage value.The lease payment is determined by on a depreciation factor, plus interest and death loss.

"Theoretically with a four-year lease and a four-year turnover rate, at the end of the four years, the leased cows would have been replaced by offspring of the herd," Hufferd explains. "The younger party doesn't have to borrow any money for the cattle." He still pays income tax but it's not all up front.

A business continuation agreement has to be combined with good estate and tax planning and with the older farmer's will. It's a complex process that you won't be able to do on your own. Because tax laws are constantly changing, you'll need good legal advice.

It helps to know what you want to do before contacting an attorney. Most attorneys are not going to tell you what's best, Ulring says. Generally, they only can lay out your options and allow you to decide.

"It's best to hire an experienced attorney," he says. Finding one can be difficult. Generally, lawyers who have experience helping farms form corporations would be best, he says, even if your farm is a partnership or a proprietorship. "Corporate attorneys draw up more buy/sell agreements than anyone else," he says.

CHAPTER 17

Don and Ruth Lowenstein and their children, one of their reasons for choosing life on a farm. Photo by John Schultz.

Technique, Not Technology: Skill and Intelligence Can Compete with Money

Businesses that never change may be en route to failure. The demise of a business can take decades, even generations. But one of the great principles of capitalism in Western civilization has been that a business must at least keep up with the changes of its competitors to survive. As we've seen in Chapter 2, farming is hardly isolated from the industrial and technological environment of our civilization. If anything, farmers may embrace technological change faster than their city cousins.

As wonderful as technology has been for farming—by making the profession cleaner, less backbreaking and, in some ways, safer—it has put the occupation on a treadmill that seems to make getting started in farming ever more expensive and exclusive.

Sometimes it takes a fresh point of view to see whether this technological treadmill is helping or hurting your farm. We've seen how Steve Hopkins and Sara Andreasen used what Steve calls "high technique, not high technology" to get started in dairying. Steve didn't grow up on a dairy farm, so he was open to new ideas about grazing and pasture management that came from New Zealand. Because Steve is willing to work hard moving fence to rotate milk cows from one small paddock to the next, he can produce milk with less investment in land and machinery than conventional dairy farms. Steve's use of this "technique" does, in fact, depend on some relatively new technology—modern electronic components that improve the effectiveness of cheap electric fencing. Rotating pastures isn't a new idea in itself. Using old-fashioned barbed wire and fence posts is an expensive way to divide pastures into small paddocks. The first electric fences were cheaper but they didn't work very well, often shorting out on tall weeds. Not until fences came along that deliver high voltage in microsecond bursts, was there a system that was safe, effective and inexpensive.

Two other smart, young farm families have also shown how agriculture can be a profitable business without making huge investments in land, machinery, and buildings. Don and Ruth Lowenstein of Cameron, Missouri, have some things in common with Steve and Sara. They didn't grow up on farms, either. Don's main occupation is accounting. And it's his eye for business practices that are losing money that helped turn an expensive hobby into a profitable sideline cow-calf business. Ruth's interest in more natural foods and farming techniques led the Lowensteins to the same controlled grazing technology used on Steve and Sara's dairy farm. Not only have the Lowensteins avoided more traditional production methods, they're also bypassing the conventional marketing system by selling beef and veal directly to consumers.

The other farmer who shows how a fresh point of view is weeding out costly technology is Tom Frantzen of New Hampton, Iowa. Frantzen grew up on a farm. He, too, has been willing to question the status quo in farming. He is a member and past president of a group of mostly younger farmers who conduct scientific tests on their farms to find less costly ways of growing crops. The group, Practical Farmers of Iowa, works closely with Iowa State University to make sure its methods are scientifically valid but farmers in the group choose the experiments. Among other things, Frantzen has tried ridge tillage and planting crops in narrow strips on his farm. He also has used some controlled grazing techniques in hog production.

Since the Lowenstein farm has only one enterprise, grass-fed beef production, and isn't as complicated as the Frantzen farm, let's look at their business first.

Don Lowenstein calls his cows, which become very tame under a management-intensive grazing system. Photo by John Schultz.

An Accountant Becomes a Cost-conscious Cowboy

On a warm day in early summer 1994, Don Lowenstein leans out of his pickup truck to yell at 40 cows and their calves at the other end of a small pasture. "Come on cows. Come aaahhn cows," he shouts, drawing it out as if he had been calling cows all of his life. Three of his six children watch from the pickup bed as the Simmental-Angus cows lope toward them. He climbs out to unhook a strand of electric fence wire. The cows rush into the next 10-acre paddock to graze a lush mix of grasses and clover on the rolling 206-acre farm near Cameron, Missouri.

Lowenstein isn't yet a full-time farmer, though he would like to be. The accountant, who is originally from Chicago, writes customized computer software for banks, public utilities and small manufacturers. He had never been on a farm until 1986, when he and his wife, Ruth, bought a house and 50 acres an hour's drive from Kansas City, where Ruth grew up.

Not until a hobby of running a few cows in the pasture turned expensive did Don cast his accountant's eye toward cutting costs. It was Ruth who found the method to do that. For Christmas 1991 she gave Don the classic book on grazing management, *Grass Productivity* by Andre Voisin. The French biochemist who developed a system of rotating cows through pastures on his farm in Normandy published the book in 1959. Ruth discovered it in magazine articles about controlled grazing.

"I guess I was kind of forced into it," Don says. "And now, in retrospect, I'm so glad that she got me into that." Controlled grazing, which is also called rotational grazing and management-intensive grazing, involves moving cows from one paddock to the next after giving the cows a short time to graze that paddock heavily. The system has allowed the Lowensteins to cut their previous bills for fertilizer for the grass and for hay purchased off the farm.

Don doesn't claim to be an expert, although he has worked hard at grazing management, including attending a grazing school at the University of Missouri's Forage Systems Research Center near Linneus.

The couple's main reason for settling in Missouri was to escape the drive-by shootings and gang graffiti invading the Chicago neighborhood where Don's grandmother had lived. "We wanted a better quality of life for our kids than we could have in the city," Ruth explains.

They picked Missouri to be close to Ruth's family. They decided to live on a farm because Ruth had fond childhood memories of her uncles' farms in Nebraska. They sold their house in Chicago in a rising market, profiting enough to buy a house and a small amount of farmland for about $50,000.

At first, "we really didn't have a clue as to what was going on," Don says. "We went through the first few years in just one-hundred percent on-the-job training." After a year they owned four cows and their calves. "Like our neighbors that owned a lot of machinery, I felt a burning desire that I needed to own machinery," he adds. But without any experience using and maintaining his growing line of used farm machinery, Don soon wrecked most of it.

In 1989, a neighboring 80 acres came up for sale and they decided to buy it—this time with a loan. That made their part-time farming venture more serious. And even before that, their annual operating costs had been doubling every year—from $1,500 in 1986 to $3,000 the next year, to $5,000 the next. In 1989, with payments on land added in, costs jumped to $12,000.

It was time to take raising cattle seriously. "We started to look at it more as a business than a big hobby," Don recalls. "And as we started to put the numbers down on paper, four things jumped out at me. One was the cost of machinery that I'd bought. Another was the amount of money that

I was dumping into feed. Another was the amount of money I poured into fertilizer. And last, but not least, was the amount of interest."

Using the controlled grazing techniques they learned at the University of Missouri's grazing school, and through trial and error, the Lowensteins cut their production costs significantly over the next several years. These are the most important changes they made on their farm:

• Don quit putting up hay, the "crop" that's typical for the Lowenstein's northern Missouri area. That immediately reduced his need for so much expensive farm machinery. Now he buys any extra hay required for the cows and he hires someone else to bale the excess grass in the spring. With the increased grass production that comes from controlled grazing, most of the feed for the family's cows now comes from rotating them into twenty 2½- to-10-acre paddocks on the remaining 120 acres of grass. The cows graze a paddock two to five days. Paddocks then rest to allow the plants to rebuild root reserves and grow more forage. The resting period ranges from 14 days in the spring to as much as 50 to 60 days by fall, when plant growth is slowing down. Even after the growing season ends, the grass stores nutrients in the field. In recent years the grazing season has lasted from early April to February. The extended grazing season and the increased grass production from controlled grazing has cut the farm's use of hay in half, from 150 big round bales to 70 or 80.

• He quit applying the nitrogen fertilizer that neighbors and his local elevator recommended for big forage yields. At the grazing school he learned to use phosphorous and potassium fertilizers to spur the growth of legumes, plants that are able to convert atmospheric nitrogen to nitrogen fertilizer. He made a one-time application of those two elements over the entire farm and now fertilizes annually only on the hay field. His yearly fertilizer bill has dropped from about $20 an acre when he followed his neighbors' methods to $6 an acre.

• He sold off most of his machinery and quit buying more. "The cows are happy to harvest the grass," he says, adding that, if he hadn't bought so much machinery, "I could have had twice as many cows by now."

• He was able to eliminate the farm's short-term debt, since he no longer had a lot of machinery and spent less on fertilizer. The farm's earnings were then used to retire land debt, which the family planned to have paid off by about 1998.

These changes in production and management cut the operating costs of their part-time cattle business from a peak of $26,000 in 1992 to about

$9,500 in 1994 (excluding capital outlays for expanding their herd and building farm ponds to water their cows).

Cutting costs is only one part of keeping a business efficient. Efficient production is important, and the grazing system doesn't sacrifice efficiency. And last, getting a market price that covers costs, depreciation and time spent on labor and management is essential. Making a little profit on top of that would make the business seem like a good investment. Marketing is the other strength of the Lowenstein farm.

"A woman living down the road from us told me farmers really have it bad. It's the only business where you have to take the price that's given to you. I wasn't convinced that was true," Don recalls. "Then I heard another farmer talk about his grass-fed chickens and beef. A light went off in my head. We don't use hormones. There are no additives in the feed. Every animal is born and raised on the farm and slaughtered right off the farm. All the animals are content. I thought, 'Someone's going to pay a premium for this.'"

The beef they sell is basically grass-fed. The steers and heifers get a small amount of free-choice corn about two months before slaughter. And, because they're all slaughtered around November 1, the yearlings that were born in different months range in weight from 850 to 1,150 pounds. Before selling any directly to consumers, the Lowensteins tried eating the beef from their own herd first, to make sure that they had a good product to offer. It was lean, with only about an eighth of an inch of fat around each cut, and it was tender. "When we first tried it, it was some of the best meat we had ever tasted," Don recalls.

The next step was marketing. "It was very simple. I started calling friends, relatives, then business acquaintances," he says. "I told them we had all natural beef with no hormones, no antibiotics; that we have small, medium and large."

They've had no trouble selling the beef, which the consumer buys live and is responsible for slaughtering and packaging. In 1994, a year when conventional cattle feeders took serious losses in spring and early summer, the Lowensteins were getting $1.60 a pound hanging weight (or 98 cents a pound live weight). Market prices that year ranged from about 77 cents a pound to 62 cents a pound live weight and cattle feedlots were losing $100 per animal or more. In 1994, the Lowensteins projected revenue of $19,560 from the sale of calves and culled cows. After deducting their $9,500 operating costs, they netted $10,000 after taxes and interest, or $62.50 per acre on the 160 acres they had in production at the time.

Obviously, that's not enough income to support Don, Ruth and their six young children. But the farm isn't at full production, either. Eventually, they could nearly double the size of their 40-cow herd. And they haven't yet established productive grass on all of the farmed-out land they've purchased. When those goals are achieved, Don may be able to spend more of his time farming.

"There's more peace of mind," he says. "The cows don't call you up saying, 'I've got an error on my hard drive.' I'm not an expert in farming; I'm a rookie, but I'm not afraid to try anything." And, unlike other businesses, he's found that learning to farm is a long-term project, because his mistakes or successes may not show up for a year. "The challenging stuff I find here on the farm," he says. "When it rains for 12 days and the mud is knee deep—that's a real life challenge."

A Traditional Farm Enriched by Innovation

The northern Iowa farm of Tom and Irene Frantzen has little in common with the Lowensteins' grassland farm of northern Missouri. In many ways it's more typical of the mixed crop and livestock farms that once dominated the Midwestern Corn Belt. The Frantzens grew up on farms. They raise cattle, hogs, corn, soybeans, forage and a recently rediscovered food crop developed by the Aztec people of Mexico—grain amaranth.

But, like the Lowensteins, Tom and Irene farm with methods not used by most of their neighbors. Tom is an experimenter, a thinker, and sometimes an outspoken critic of the agricultural establishment. Over the years he has tried planting crops in narrow strips, a practice which helps stop soil erosion and may also provide some yield advantages, especially in areas of the country with marginal rainfall. He has also tried controlled grazing techniques for both his cattle and for gestating sows. And he plants his crops using "ridge tillage," a conservation tillage technique that saves soil from erosion almost as well as using no tillage at all.

Ridge tillage involves building up a ridge of soil in the row of a growing crop of corn or soybeans in the summer. During each cultivation, soil is thrown into the row, burying any small weeds. The following spring, a special ridge-till planter is used to level off the tops of the ridges to create a seedbed. The process also throws the old plant residue from the previous year's crops—cornstalks and leaves, for example—back in between the rows. That acts as a mulch that slows down germination of weed seeds between the rows. Among the pioneers of this ingenious system were researchers at the University of Nebraska and Fleischer Manufacturing of Columbus, Nebraska, which made equipment designed to create a seedbed on ridges—the Buffalo Till Planter. Fleischer and several other companies now make ridge-till planters and heavy-duty cultivators that work in the residue-filled spaces between rows.

Ridge tillage has become more popular as conservation requirements for federal farm programs have become tougher in recent years, but it has not caught on as well as no-till. That's another system that uses no tillage at all. It requires less machinery and less fuel than ridge tillage. No-till uses about the same amount of herbicide as more conventional farming, which relies on some form of plowing or disking to prepare a field for planting. No-till's advantage over these conventional tillage methods is that making fewer trips over a field saves fuel. No-till also allows operators of large farms to plant their crops quickly. If the herbicides are effective, no-till saves cultivation time. Unfortunately, for all of its advantages, no-till may also be the latest belt on the treadmill of technology. Because it frees up a farmer's time, it has added incentive for large cash grain farms to outbid others for rented cropland. This may be one of many factors pushing rates for rented land higher, and out of reach for many beginning farmers.

For smaller farms, like the Frantzens,' ridge-till offers the advantage of using fewer chemicals than either no-till or conventional tillage because the farmer can apply both fertilizer and herbicides only in the raised seedbed on top of the ridge, not over the entire field. It takes a little more time and fuel for cultivation, however.

Both ridge-till and no-till are crop production systems that have merit for young farmers. Ridge-till may take a bigger capital outlay for cultivating equipment than no-till, but no-till likely will require higher annual operating expenses for chemicals. Both systems appear to be more profitable than conventional tillage and it's about a tossup as to whether ridge-till or no-till is the most profitable. At least that seems to be a logical conclusion to draw from a program sponsored by *Successful Farming* magazine called the MAX, which stands for "Farming for Maximum Efficiency." The program, which was also run by the national Conservation Technology Information Center and two farm chemical companies, has shown that both systems do well. The program compiles records of production costs and crop yields from farms using a variety of tillage systems—500 farms in 1993—and ranks them by profitability. It also adds a charge for soil erosion—$5 for each ton of soil lost above acceptable limits from an acre. That's not a cost that would show up in normal business accounting but it could show up in the future in the form of reduced yields. The program has shown that ridge-till has about $10 an acre lower chemical costs than other systems. But field operation costs tend to offset that advantage. Yields from both systems are similar.

In 1993, a ridge-tiller had the most profitable corn following soybeans among all Iowa farms in the MAX program, with a net return of $242 an acre. A no-tiller ranked third, with a profit of $180 per acre. But that was an unusually wet year in Iowa. Illinois, which had more typical weather over much of that state, showed a less dramatic difference. A no-tiller ranked first in soybean profitability, with a net return of $143.62 an acre and a ridge-till farmer was a close second with a $142.24 an acre profit.

For the Frantzens, ridge-tillage has been a good choice. They farm 320 acres—less than the average size for Iowa—and are raising three children with no off-farm income. Tom thinks it makes more sense to use his labor on the farm than to buy a lot of off-farm supplies, including chemicals.

"We want to put our efforts first and purchase what inputs we need second and try to make a profit off of that," Tom says. "The goal here is not the highest yield. The goal is to make a profit."

Frantzen adds that they don't follow an ideology about their methods, either. Although Frantzen's group, Practical Farmers of Iowa, is sympathetic to organic farming, most of its members' farms aren't completely organic. The Frantzens' farm isn't, either. To see how ridge-till works, let's look in detail at the Frantzens' soybean enterprise. "It's as good a moneymaker as we have because we don't spend any money on them," he says.

That's a slight exaggeration of course, but his records show less spent than by members of the Southeast Minnesota Farm Business Management Association, for example. The Frantzens' farm is not far from the Minnesota border and is similar to the farms in the Minnesota group. The Frantzens' direct—or variable—costs (excluding land and other overhead expenses) come to $69 an acre, compared to the $86.43 an acre average for the Minnesota group in 1993. Iowa State University estimates total variable costs for raising soybeans (using conventional tillage) at $90.37 an acre in 1993.

"We save an easy $20 to $30 an acre by ridge tilling," Tom Frantzen says.

Their biggest savings is in herbicide expenses. Frantzen spends about $7 an acre on herbicides, mainly Assure II, a grass herbicide used after the soybeans emerge from the soil, and some Roundup and 2,4-D. The average for the Minnesota group in 1993 was $24.74. The Frantzens also clean and bag their own soybean seed, which cuts their seed cost to about $10 an acre, versus $13.33 an acre for the Minnesota group.

Another enterprise that's more complex but also efficient is the 80-sow farrow-to-finish swine herd. Many aspects of the business are fairly typical. For genetics, Frantzen uses a traditional Hampshire-Yorkshire-Duroc crossbreeding program. The market hogs are fed to 125 pounds in a standard slatted-floor grower building and then to slaughter weight (240 pounds) in outdoor Cargill-type cement-lot finishing units. The Frantzens sell about 1,200 hogs a year. Their break-even cost was about $31 per hundredweight, not low enough to be profitable during the financial blood-bath the industry suffered in the fall of 1994 but well below the costs of many huge corporate farms that year. The quality of the Frantzens' hogs was better than average, with backfat running between .8 inch and .9 inch and 49 to 50 percent of the carcass being lean meat.

"We're good, but we can't get the industrial cookie cutter," Frantzen says, referring to the very lean look-alike hogs that come out of corporate hog factories. "But I don't believe the industrial cookie cutter works in sustainable agriculture." And those industrial systems have to get premiums for top-quality hogs to make their more-expensive production systems profitable, Frantzen contends.

But the rest of the Franztens' system has some strong advantages that make it look "sustainable" both from an ecological and financial point of view. It's in the gestation and farrowing stage of his hog production that the Frantzen farm eliminates many costs common to both the largest hog factories and to many medium-sized independent operations that put hogs in buildings during the entire production process.

During gestation, the farm's sows graze in 30 half-acre paddocks alongside cattle. That phase of the hog system works about the same as the controlled grazing used for beef production by the Lowensteins in Missouri—with the added advantage that cattle and hogs prefer different plants in a pasture, making the use of forage even more efficient.

When the sows are ready to farrow, they're moved to huts in a pasture system that may be unique to the Frantzen farm. It combines crop rotations, modern controlled grazing techniques, and old-fashioned "hogging down" of corn—in other words, letting hogs eat a mature crop of corn in the field instead of harvesting, drying and storing the corn.

Frantzen has divided the 15-acre farrowing area into long strips of red clover pasture, oats and corn. There are six sets of three strips, with each strip planted to one of the three crops. Each set of three strips is also protected by a planting of trees or bushes that will become a windbreak when mature. Between these windbreaks, the three crops are planted in a three-year rotation. Oats and red clover are planted in a strip one year. Frantzen harvests the oats by machine in early summer, leaving the clover to form a thick stand. The next year, he puts farrowing huts in the strip of clover pasture, where the sows graze. The following year, that strip is planted to corn. In the fall, sows and pigs are turned into a corn strip that was planted the year before. It has 16 rows of corn that are 500 feet long. In other words, each set of three strips always has one strip of pasture available for farrowing, another strip of oats where a young stand of clover is developing for the following year, and a crop of corn that's maturing for fall harvest by the sows that are in the pasture strip.

Frantzen uses electric fencing to keep the sows in the pasture, to keep them off the oats, where they would damage the young clover, and to time when he turns the sows into the corn. The sows do most of the eating and the pigs gain weight by nursing. The pigs are weaned at four to six weeks, weighing about 40 pounds.

"It's an enormously big pig to be weaning," Frantzen says, laughing. "But the sows are eating the corn, which I am not going to complain about." Letting the sows do the work may save as much as $60 an acre, he estimates, because he doesn't have the expense of combining corn, drying it, grinding it into feed and hauling manure from farrowing and nursery buildings back to the field.

Many hog farmers might view this system as wasteful, since there's no control over how much the sows eat. And the common wisdom is that pasture farrowing takes more labor than modern, mechanized farrowing buildings and nurseries. In 1994, Frantzen conducted a research project that seemed to disprove both criticisms and showed that his system might be very competitive. With the help of a $1,500 federal Sustainable Agriculture Research and Education (SARE) producer grant, Frantzen bought a scale that allowed him to weigh the sows before farrowing and at weaning and to weigh the weaned pigs. He also kept track of additional purchased feed for the sows and pigs.

Here's how he described the test in his preliminary report to Iowa State University:

> On August 13, 1994, I weighed seven sows and hauled them to a narrow strip of red clover pasture. They began farrowing on August 24th and finished on September 4th. On September 14th, I removed one of the electric fences and gave them access to an adjacent strip of 103-day corn. The corn was completely eaten by October 13th. On October 17th,

all seven sows and their 60 pigs were loaded on a cart, weight checked and hauled home. Following is a summary of the weights, the feed used, and the hours of labor involved:

Weights

August 13th . . . . seven sows=3,110 pounds incoming weight

October 13th . . . . seven sows=2,480 pounds outgoing weight

October 13th . . . . 60 pigs=2,410 pounds weaning weight

Feed Used

1,515 pounds of dry corn at $1.90 per bushel $53
   70 pounds of sow premix 21
   100 pounds of soybean meal 9
   420 pounds of sow cubes 63
   150 pounds of hog mineral 48
   775 pounds of pig starter for young pigs 184
   425 pounds of pig starter 58

.6 acres of clover pasture 34.50
.6 acres of standing corn 108.60
(Clover and pasture costs are based on an $85/acre land charge, $96/acre cost to grow corn without harvesting and $30/acre cost to establish clover. The clover cost was divided in half because there are at least two farrowings per season on each pasture.)

Labor

set and remove fence . . . . .0.4 hour
repairs . . . . .0.3 hour
hauling stock . . . . .1 hour
process pigs . . . . .0.7 hour
chore time . . . . .6.2 hours
Total hours of labor   8.6 hours @ $10/hour. $86

Total costs $690.60

If all of these costs are allocated to the pigs, their cost comes out to

$11.09 per pig. Add in the cost of maintaining each sow for one gestation (about $25 on the Frantzens' farm) and the cost of raising feeder pigs with this system comes to about $14 per pig. That's less than half the industry's estimated break-even price of $30 for a 40-pound feeder pig in the late fall of 1994. (This analysis does not include Frantzen's cost of replacement sows, which in Frantzen's system would be negligible.)

Not only is the cost of production very competitive, the amount of time involved in this system compares well with the most mechanized systems. "One of the things I want to dispel is that pasture farrowing is labor-intensive. That's bull," Tom Frantzen says. "There's no manure to haul. There's no cleanup."

Frantzen's system might work well for young farmers without a lot of capital, but it also has some disadvantages. One is that, in northern Iowa, cold weather limits the outdoor farrowing season to May through September. If farrowing is tried in October, "I've seen disasters—terrible storms. You get poor pig performance," he says.

In early summer, though, the growing strips of corn create a warm microclimate for the clover pastures sandwiched in between. Even on windy days, "everybody's out there grazing," he says.

So far, Frantzen has had no trouble finding a place to sell his hogs but his biggest worry is that he will eventually have trouble finding good replacement hogs. The genetics for mechanized indoor production won't work as well outside. Frantzen selects his replacement gilts from the sows that thrive on pasture and that do a good job of "hogging down" corn. "If the genetics are lost, this becomes a terrible risk," he says.

Most observers of the rapidly changing hog industry expect some independent producers to survive, which may keep the genetics for the Frantzens' system available. As we'll see in the next chapter, hog farmers are already forming informal groups, corporations and co-ops to work together to compete with large companies. Having low production costs will be essential in any sector of agriculture in the future, but that alone may not be enough for independent farmers to survive.

CHAPTER 18

A Paradox for Family Farms: Cooperate to Stay Independent

Fred Heigele and Perrie Seifert (*left*) sort hogs for shipment to market by a trucking co-op in Kansas. Photo by Ed Lallo.

In late November of 1994, when nearly everyone in the country who produces hogs was losing money, Tom Meek of Clay Center, Kansas was losing a little bit less.

"I haven't gotten a check under 30 cents a pound yet," Meek said, at a time when most packing plants were paying farmers 28 to 29 cents and in some areas prices were as low as 26 cents. Meek wasn't exactly celebrating. Meek, 42, buys feeder pigs and finishes about 1,300 a year. At the time, his break-even price was 35 cents a pound. "I'm basically losing $10 a hog. You can figure I'm donating my labor."

Yet, at a time when the industry seemed to be consolidating into the hands of large companies, Meek and 14 other north central Kansas farmers were surviving, thanks to an informal marketing group put together by a neighbor, Roy Henry. Henry, who became Kansas Pork Producers Association president late in 1994, is a member of a family corporation that runs a 900-acre farm and a swine enterprise that sells 16,000 hogs annually, including 6,000 gilts for a multiplier herd of PIC (formerly Pig Improvement Company).

A decade earlier, Henry started the group when his own hog business had grown "to the size that we needed a semi" for shipping hogs to market—instead of taking smaller trailer or pickup loads to town. He began shipping direct to a Farmland Industries packing plant in Crete, Nebraska, 100 miles to the north. At the time, Henry had only enough hogs to fill half of a semitrailer and he recruited other neighbors to share space. "If it was good enough for me, it was good enough for my neighbors," Henry recalls.

By 1994, the group was shipping 40,000 hogs a year to Farmland, in several loads a week. In exchange for shipping on days when the regional co-op's packing plant most needs hogs, Farmland pays the Kansans a modest premium of 50 cents to $1 over the price at the plant. In practice, members of the group have done even better than that, in part because they have compared records that the packer gives each farmer on how much lean meat their hogs produce. The farmers have changed their production or hog breeding practices to try to raise more of the lean type of hogs the packer wants. They get higher premiums for improved quality.

The marketing group is informal. It's not a separate legal entity like some marketing groups. Henry works with the packer to pick the best marketing times and he arranges transportation. He charges 20 cents a hog for his work, which he uses to cover his phone expenses. Anything left over goes to charity. "I'm not in it for the bucks," he says. Trucking costs another 65 cents per hundredweight of hogs shipped.

Although Henry doesn't personally profit from the time he spends running the marketing group, all of the members have improved their income from hog sales through the group. Their success was confirmed by a 1993 study of marketing groups by Kansas State University economist Jim Mintert and a graduate student, Richard Tynon. The study looked at ten marketing groups in Kansas and in Iowa and found that they improved members' prices by an average of 60 cents per hundredweight. That was hardly enough to offset losses when market prices were depressed by large supplies of hogs, cattle and poultry in the mid-1990s. But because the marketing groups also saved on transportation and other costs, members increased their net returns by $1 to $1.75 per hundredweight. Privately, some groups report premiums of as much as $4.50 per hundredweight over local prices, in part due to improved quality.

In Illinois, the nation's largest hog marketing cooperative, Hog, Inc., with annual sales of $40 million, has netted producers premiums ranging from $1.50 per hundredweight to $8.69 per hundredweight, manager Kirby Bates told *Successful Farming* in 1993. The co-op based in Carlinville began as a co-op organized to buy production supplies at a discount for members.

In order to really benefit from the advantages that a co-op can give individual hog farmers, some advocates, including Roy Henry, think they need to do more than simply pool hogs for marketing in volume. Henry is trying to talk some of his neighbors into starting a farrowing group, which would provide three-week-old pigs to its members. Individual members of the group would be in charge of different stages of segregated production, "so we can use family labor to raise high-health pigs," he says. Henry also believes that having feed milled to the members' specifications might lower production costs. Some commercial feeds have added ingredients that aren't necessary, he says.

Still, some young farmers are wary of the added financial commitment required for farrowing co-ops, for example. Steve Luthi is a farmer in his 30s who ships hogs to market with Roy Henry's group. Luthi, who owns half-interest in a 225-sow farrow-to-finish operation with his parents, is pleased with the marketing group's results. But the Luthis' operation is relatively inexpensive. The sows are in dirt lots and some hogs are finished in open-front buildings. The Luthis believe that the genetics they buy and their methods result in good-yielding animals and they aren't interested in a farrowing co-op.

Tom Meek says he favors starting a farrowing co-op. "It would give me a more secure feeling if I knew I had a long-term source" of pigs, he says. But Meek is well aware of the potential cost of such a venture. In Iowa, one farrowing co-op cost its 15 members $70,000 apiece to invest in the buildings and other start-up costs.

That's a relatively small investment by the standards of commercial agriculture. But for young farmers with little capital, joining co-ops that coordinate production as well as marketing may be a difficult financial obstacle.

And even marketing groups seem to attract relatively well-established farmers. A survey by Kansas Agricultural Statistics released in 1994 showed that only 8 percent of that state's hog farmers sold through groups. Yet larger producers used groups most often; 40 percent of those selling 7,000 hogs a year or more sell through groups.

Group marketing at the very least means that members will have less control over when and where they sell their hogs. That concerns Brad Murty, an agricultural student at Iowa State University who is planning on going into hog production. "There's going to be less competition. How do you think the packers will respond? I don't think they'll pay as much."

Aside from the issue of price, the threat of losing access to packing plants as big companies begin to dominate sales to those plants was one of the reasons Randy Reifschneider of Hubbard, Iowa, helped start a marketing group in 1994. "I got frustrated. I would call Monfort (in Marshall-town, Iowa) and they'd say, 'We're full today; call back tomorrow.' And you know the market would be lower the next day," he says. Reifschneider and 24 others invested $600 apiece to form a corporation, Frontier Quality Pork, which hired veterinarian Tom Samp to seek bids from a half-dozen packers. Part of the money was used to remodel a building that Frontier uses as a collection point for loading members' hogs onto a truck.

One alternative to joining a marketing group is to farrow or finish hogs on contract with a large company. Kansas producer Meek knows of two young families in his area doing just that with Farmland Industries. Farmland is a co-op, although some independent producers consider Farmland's ownership of hogs as direct competition with them. For young farmers, contracting seems less risky because the contracting company usually bears the market risk. The disadvantage is that some contracts, especially for finishing hogs, often pay just enough to pay off a loan for putting up a new building and pay little for the young farmer's labor, Meek says.

Contracting is feared in much of the Midwestern Hog Belt, where farmers see it as just one step to complete vertical integration, where a few companies control all of the production, from breeding replacement animals to processing and marketing to consumers. That is exactly what has happened in the poultry industry. Today, something between 10 and 20 percent of the swine industry is vertically integrated. If the industry follows the poultry industry, independent producers will have a tough time finding packers willing to buy their hogs.

Yet, even in the poultry industry, young producers sometimes get started on their own with bank loans requiring 0 to 5 percent down, says Henry Brandt, senior vice president at The People's Bank of Maryland in Denton. An integrated livestock industry may not be as profitable as independent production, but even in the poultry industry, some opportunity remains for enterprising young farmers.

Whether young hog producers work with contractors or in some kind of producer-controlled business or co-op, Meek thinks the industry is changing so much that traditional "independence" will vanish in the hog industry.

"I think we're going to have to look at some new directions," he says. "I think it's going to be harder to start out farrowing 20 sows like we did a few years ago," he says. Meek and his wife, Nancy, have two sons under 12 who haven't yet decided on a career. If a young person doesn't want to contract, "I think some kind of marketing group is crucial," he says. "The fact that the margins are a lot tighter makes any extra profit all the more important. If my son was wanting to start farming now and needed to raise hogs, I think I'd advise him to try to find somebody's coattails to jump onto."

In spite of the pessimism that permeated much of the hog industry in the mid-90s, cooperatives may still offer the best means for independent farmers to share in the good times that inevitably will return to the industry.

Hog, Inc., for example, welcomes small- and medium-sized producers and has had members who sell as few as 16 hogs a week, says manager Kirby Bates. All members pay 40 cents a head to sell through the group, with a $300 minimum commission. The co-op charges an extra 5 cents per hundredweight for each additional stop at a farm to fill a truck.

"Everyone has an equal say in the co-op," Bates says. "And we've found that one of our strengths is in the sharing of ideas from members of all sizes."

So far, few independent producers have chosen to join co-ops that market, or that coordinate some stages of production, or that buy supplies together. But that may be changing. The feeder pig co-op that Dwight and Sally Puttmann of Kingsley, Iowa, helped form was looking into group marketing. Al Rudin, a hog farmer at Conroy, Iowa, at the other end of the state supported an effort to start a feeder-pig co-op that, so far, hasn't gotten off the ground. Rudin says he was disappointed. If independent farmers don't cooperate, he says, "you're just handing the business to your competitors without even fighting."

Lessons from Resourceful Pioneers

Dan Looker interviewing Don Lowenstein for this book. Photo by John Schultz, courtesy of *Successful Farming*.

When I set out to write this book, I did not intend to write a step-by-step manual that would tell you exactly how to get started in farming.

There was a time when late-night television offered shows on how to make a killing in the real estate market. When the residential real estate market collapsed in California and other populous states, the folly of believing those television shows became obvious. It would be almost as deceptive, I think, to suggest that any single book could ever tell everyone exactly how to start farming. Every region, even every farm is a little different. Agriculture is so diverse, and so complex, that no single source of information can get you started.

What I hope this book had done, instead, is to offer some inspiration and a few themes, a few threads that tie together the different experiences of successful start-up farms and the well-planned business arrangements between older farmers and their younger successors.

If there is a single message here, it's that getting started in farming today is still possible—but it's not easy. For most young people, farming means having less leisure time, less security, fewer benefits and often less income than their city friends with a job. In the long run, though, the beginning farmers who become tomorrow's established business owners will have more independence, more control over their own lives and, perhaps, more real wealth. Farming is still a way of life, too, as well as a business. It's one that can bring families closer together and can foster the self-re-

173

liance and discipline that helped build this nation. In short, it fosters the real "family values" that politicians like to talk about. I don't want to romanticize this. Anyone who has grown up on a farm or in a small town knows that rural America has its share of unhappy and maladjusted people. But farming has to be a healthier environment for people than many urban neighborhoods, both rich and poor.

Virginia produce farmer Chip Planck, featured in the chapter on apprenticeships, made this point eloquently when he spoke to the Pennsylvania Sustainable Agriculture Association in February 1995: "The best thing we could do for the world, not just for America but for the world, with its unlivable megacities, would be to point the way back (or forward) to commercially viable and environmentally cautious farming. Whether people are drawn to cities by high wages and ease, or driven to them by rural decline and exploitation, the results are the same: Too many people living cheek by jowl, unable any longer to take care of themselves."

Beyond showing that it's still possible to start farming, a few themes stand out from my interviews from this book, and from helping *Successful Farming* magazine organize two national conferences for beginning farmers and ranchers.

The first lesson has to be that a beginning farmer's survival is enhanced by a good education. That doesn't mean too much book learning. Moe Russell of Farm Credit Services of the Midlands in Omaha, Nebraska, recommends Kirkwood Community College in Cedar Rapids, Iowa—a two-year school. Agriculture is constantly changing. If you learn to navigate the Internet with a computer, or to do a thorough "literature search" at your land-grant university agricultural library, then you'll know how to find state and national experts on, let's say, swine nutrition or soil science. That skill will help you long after the books you read in college are obsolete.

A second lesson is that education also has to mean more than a formal education from books. Joe Hlas, the young Iowa State University student recruited to work on the farm of Dwight and Sally Puttmann of Kingsley, Iowa, had a nearly straight-A grade average in school. Yet, Joe says his real education began when he worked with Dwight. At the Puttmann farm, Joe learned how to farrow sows, how to manage his time, and how to put practical skills in carpentry and electrical wiring to work. He grew up on a farm, but that farm fed feeder pigs to market weight. It didn't farrow sows.

Chip Planck, who was a college professor before he became a farmer, puts it well: "The best ag school of all is your own farm. The next best, working at another farm for money. The next, working at another farm for no money. The worst, an ag school where you must pay them."

Planck adds that you'll learn the most from your own farm if you work it full time. Although off-farm income is crucial for most start-up farmers, Planck advocates quitting that job as soon as possible. "The demands of non-farm work will always conflict with crucial moments in a growing season. Being unready for a few of those points each year—to plow when it's momentarily dry in a wet spring, to keep a greenhouse swept free of snow so it won't collapse, to pick your peppers out before a hard frost—can forever keep you from becoming self sustaining."

When Planck spoke to aspiring farmers in Pennsylvania, his audience included many who didn't grow up on farms. He advised them to respect and learn from traditional farmers. "There is no 'organic' way to pick beans," he said. "There is no 'bio-dynamic' strategy for adjusting carburetors or chiseling rusted bolts from the worn cultivators you got for a steal at an auction." His advice was to read farm magazines and books, get to know your farming neighbors, go to auctions, and "take welding or auto repair at the high school before taking agronomy at Penn State or deep digging at UC Santa Cruz."

Just as would-be farmers from the city can easily romanticize and underestimate the skills and intelligence that agriculture demands, there is another trap that faces those who grew up on a farm. The children of farmers who believe they will be able to farm like their parents may be doomed to fail. If you want to be an independent hog farmer, maybe you should go to work for Murphy Family Farms or Premium Standard Farms, to see what they do right and what they do wrong. If you want to grow corn and soybeans, maybe you should work for a California tomato farmer or a Washington state apple grower. Many fruit and vegetable farmers sell their crops without any help from a government farm program. The odds are high that corn and wheat farmers will soon be selling their crops with little help from the government, too. The parents who raise wheat and corn now have spent a lifetime planning their crops with the farm program in mind. For a lot of reasons, a young person could benefit from experience outside the farm where he or she grew up.

For any beginning farmer, the real world has to be your college, whether you grew up on a farm or not.

A third theme runs through this book—frugality and thriftiness. Most of the young farmers you've met are very cautious about borrowing these days. Some would say farmers are downright cheap. They are, and there are good reasons for it, like drought and crop insurance that doesn't really cover all losses, like uncertain interest rates and the memory of neighbors who borrowed too much and now live in town. Even though credit is a necessary tool for commercial agriculture, you cannot borrow your way to wealth.

Closely related to the issue of debt and money management is the need for any young farmer to have good business planning skills—an understanding of basic bookkeeping, financial analysis and strategic planning. If you're going to raise cattle on a ranch today, you need to know how to push a pencil—or computer mouse—as well as how to pull a calf from a heifer giving birth for the first time.

For those reading this book with no intention of farming but who want to help beginning farmers, we've already seen that state governments or state-level organizations can help many young farmers for very little money by running a linking program to match young people with farmers nearing retirement. And we've seen that rural statesmen like former Congressman Cooper Evans, have good ideas for helping young farmers with relatively inexpensive federal programs.

Steve Hopkins, one of the smartest young farmers I know, shares some of Evans' ideas for federal help. He, too, would like to see a capital gains tax break just for older farmers who sell to a beginning farmer. "I've encountered so many older farmers who are hanging on to their farms because they can't bear that tax burden," he says.

Hopkins would also like to see tax-loss farming outlawed, even distinguishing between hobby farmers and those who try to make most of their living from farming. And he'd like to see young farmers given a chance to buy or use land in the Conservation Reserve Program—the ten-year land retirement program that the federal government used to take erosion-prone land out of crop production. Hopkins' type of grassland farming wouldn't harm the land. If anything, it would improve it.

I'd like to see those things done, too. But I'm not optimistic. The federal government, regardless of which political party is in power, is likely to be in a mood of retrenchment and cutting programs because of the crippling national debt. That atmosphere seems to be discouraging innovation. Programs to help young farmers don't have to be expensive. But such a small constituency is likely to be ignored in Washington.

I'm a little more optimistic about good ideas continuing to come from farm states and local governments. Hopkins, again, has some good ones. He thinks rural communities ought to offer incentives to young people to buy run-down farmsteads. A rural utility has already offered incentives to buy vacant houses in northeast Iowa, he says. And chambers of commerce will bend over backwards to attract a Wal-Mart or small industry. Similar incentives to young farmers would be just as good for the local economy, maybe better.

The county Extension office ought to have a "beginning farmer packet" that tells new people in the community where to get LP gas for their home, where they can buy hog feed, for example. "It would be kind of like a Welcome Wagon" for new farmers, he says. Local farm stores, too, ought to offer a discount to new farmers in the community, or perhaps a gift certificate.

Eventually, I think that ideas like those of Steve Hopkins will take root in rural America, because leaders there know that keeping more farmers in the community is one way to slow population decline.

Another reason that I'm less optimistic about help from the federal government is that some people believe that the government doesn't need to help young people start farming. One of them is Luther Tweeten, a respected agricultural economist at Ohio State University. In an article in *Choices* magazine in 1994, Tweeten pointed out that there is, indeed, more opportunity in farming in the 1990s than in the late 1950s. He compared the rate of retirement and death of farm operators with the rate of farm-raised males reaching the age of 25 (the traditional, if sexist, pool of new farmers). He found that in 1959, only 80 percent of those young men from farms could expect to replace older farmers. By 1987, he wrote, the ratio had reached 115 percent—suggesting a potential shortage. Tweeten argued, though, that most of the nation's food is grown by some 300,000 commercial farms and that only about 5,000 new farmers a year are needed to keep those farms going. Tweeten believes that small number can be met by the sons and daughters of existing farms and the "nontraditional operators" that might include farmers like Chip Planck and Steve Hopkins.

There's little reason to doubt Tweeten's analysis of the demand, but to suggest no need for any federal intervention seems curious even if the demand is potentially small. Let's look at my profession—agricultural journalism. As the number of farmers has declined, the potential readership of farm magazines and newspaper coverage of farm issues has shrunk, too. (*Successful Farming*'s readership has stabilized and grown slightly in recent years because the magazine has been able to expand readership outside of its traditional area of strength in the Corn Belt.) If a dozen or more farm writing jobs open up at newspapers and magazines in a year, that's a lot. Yet, most universities with agricultural colleges are training far more people than that in agricultural journalism. And, at least for now, those students can get subsidized college loans to study for a career that's virtually closed to them.

Tweeten's critique is in keeping with his willingness to take a hard look at many of the programs and tax breaks that the federal government has extended to farmers for half a century. He has been courageous and intellectually honest. His message that farmers as a group are now richer than most taxpayers who support farm program subsidies hasn't been popular with some farmers.

Still, I think it's in the public's interest to level the playing field for beginning farmers and to encourage new blood in agriculture by providing modest credit programs for entry-level farmers who aren't wealthy. And, if economists' projections for increased food exports from the United States do materialize, there may indeed by a shortfall of young farmers. Until the need is obvious, though, I expect Luther Tweeten's analysis to prevail in Congress.

Meanwhile, young farmers who are able to find a plot of ground to farm should remember Chip Planck's number one rule for beginning farmers:

"Put a crop in the ground this spring, tend to it like a peasant who would starve without the harvest, and sell every stick of it for as much money as you can get."

Appendix

Finding More Help and Ideas

Donald J. Jonovic, a consultant who writes *Successful Farming*'s "Can Their Problem Be Solved" column has been a popular speaker at the magazine's national conferences for beginning farmers and ranchers.

Like an Ann Landers for rural America, Jonovic deals with personal problems and strained family relationships that can destroy a successful family farm. Conflicts between generations are a common theme in his work. With humor, Jonovic describes the most typical problems and how they might be worked out. A key piece of advice he leaves with audiences is that you can't solve everything yourself. At times, you're going to need outside advice. When a listener once asked Jonovic if he knew of any good books about estate planning, he replied, "That's like coming up to me and asking, 'Jonovic, I've got this lump on my cerebellum. Know any good books about self-administered brain surgery?'" For a good estate plan, you could need the services of a lawyer, accountant and insurance agent.

With that caveat in mind, I'm going to list some books on estate planning, financial planning, marketing and other important subjects that should be considered by beginning farmers and by parents or older farmers who are helping them. As the conclusion to this book points out, success in farming tomorrow will come from a combination of solid formal education and practical experience. This book has been an introduction to beginning farming. The following list will give you more valuable background.

After the list of suggested reading, are two other important lists. The first describes linking programs to match unrelated older farmers with young people who want to farm. In the year that I worked on this book, it grew from 13 to 16. More are planned, so if you don't see your state on the list, call your state agricultural university's agricultural economics department and your state department of agriculture to see if either has a linking program, or if the staff knows of one near you. The second list describes state lending programs for new farmers. Most use "aggie bonds" and are run by state agriculture departments. If your state isn't on that list, call your ag department, or call Bruce Abbe at Communicating For Agriculture (CA) at that nonprofit group's Bloomington, Minnesota, office at 612/854-9005. CA has promoted aggie bond programs.

Suggested Reading

Family Farming: A New Economic Vision by Marty Strange, 1988, Institute for Food and Development Policy and the University of Nebraska Press at 901 N. 17th St., Lincoln, Nebraska 68588-0520. $18.95.

A hard-headed look at the forces that threaten family farms.

Farm Estate and Business Planning by Neil E. Harl, 1994, Century Communications Inc., 2821 N. Duff Ave., Ames, Iowa 50010. $24.95 plus shipping and $3 handling.

Harl, an attorney and agricultural economist at Iowa State University, is a national authority on tax law and agriculture.

Passing Down the Farm: The OTHER Farm Crisis by Donald J. Jonovic and Wayne D. Messick, 1994, Jamieson Press, P.O. Box 909, Cleveland, Ohio 44120. $24.95.

A look at typical family issues involved in estate and succession planning on farms.

Sell What You Sow: The Grower's Guide to Successful Produce Marketing by Eric Gibson, 1994, New World Publishing, 3701 Clair Drive, Carmichael, California 94608. $22.50.

An excellent beginner's manual on direct marketing.

Small Farm Handbook edited by the Small Farm Center, University of Calilfornia, Davis. 1994. Published by ANR Publications, Division of Agriculture and Natural Resources, University of California, 6701 San Pablo Ave., Oakland, California 94608-1239. $20.

A primer for beginners, with tips on buying used machinery, record keeping, raising crops and caring for livestock.

Weighing the Variables: A Guide to Ag Credit Management by David M. Kohl, 1992, Doan Agricultural Services Company and the American Bankers Association, 1120 Connecticut Ave. N.W., Washington, D.C. 20036, $60.

A readable guide to all of the financial ratios and measurements used by bankers to measure the financial health of a farm.

State Linking Programs

California
Ag Link
P.O. Box 1
Ballico, CA 95303
800/588-LINK

Illinois
Prairie Farmer
P.O. Box 3217
Decatur, IL 62524
217/877-0679

Iowa
Farm-On
Rural Concern
10861 Douglas, Suite B
Urbandale, IA 50322
800/747-5465

Kansas
Kansas Farm Link
Discontinued by the Kansas
   legislature in 1995.

Massachusetts
New England Land Link
(c/o New England Small Farm
   Institute)
Box 937
Belchertown, MA 01007
413/323-4531

Michigan
Young Farmer Department
Michigan Farm Bureau
7373 Saginaw
Lansing, MI 48909
800/292-2680

Minnesota
Minnesota Farm Connection
Passing on the Farm
1593 11th Ave.
Granite Falls, MN 56241
800/657-3247

Nebraska
Land Link
Center for Rural Affairs
P.O. Box 406
Walthill, NE 68067
402/846-5428

New York
NY Farm Net
Dept. of Ag Economics
Warren Hall
Cornell University
Ithaca, NY 14853-7801
607/255-1603

North Dakota
Farm Link
North Dakota Dept. of Agriculture
600 E. Boulevard, 6th Floor
State Capitol
Bismark, ND 58505-0020
701/328-2231

Ohio
Ohio Farm Link
Ohio Council of Churches
89 East Wilson Bridge Rd.
Columbus, OH 43085-2391
614/885-9590

Oklahoma
Oklahoma Ag Apprenticeship Program
c/o Wayne Walters
Route 1
Canute, OK 73626
405/472-3320

Pennsylvania
Pennsylvania Farm Link
Center for Rural Pennsylvania
212 Locust St., Suite 602
Harrisburg, PA 17101
800/9PA-FARM or
717/787-9555

South Dakota
Farm Link
South Dakota Dept. of Agriculture
523 E. Capitol Ave.
Pierre, SD 57501
605/773-5436

Utah
Farm & Agribusiness Transfer Program
Utah State University
UNC 4800
Logan, UT 84322
801/797-2267

Wisconsin
Farmers Assistance Program
Wisconsin Dept. of Agriculture
Trade & Consumer Protection
2811 Agriculture Drive
P.O. Box 8911
Madison, WI 53708-8911
800/942-2474

State Beginning Farmer Loan Programs

Arkansas Development Finance
    Authority
P.O. Box 8023
Little Rock, AR 72203
501/682-5900

Colorado Agricultural Development
    Authority
700 Kipling, Suite 4000
Lakewood, CO 80215-5894
303/239-4114

Illinois Farm Development Authority
427 East Monroe, Suite 201
Springfield, IL 62701
217/782-5792

Iowa Agricultural Development
    Authority
Wallace State Office Building
Des Moines, IA 50319
515/281-6444

Kansas Development Finance
    Authority
700 SW Jackson, Suite 1000
Topeka, KS 66617
913/296-6747

Minnesota Rural Finance Authority
90 W. Plato Boulevard
St. Paul, MN 55107-2094
612/297-3557

Missouri Agriculture and Small
    Business Development Authority
P.O. Box 630
Jefferson City, MO 65102
314/751-2129

Nebraska Investment Finance
    Authority
1230 "O" Street, # 200
Lincoln, NE 68508-1402
402/434-3900

The Bank of North Dakota
P.O. Box 5509
Bismark, ND 58502-5509
701/328-5672 or 800/472-2166,
    ext. 85672
(The bank has no aggie bond programs,
but administers its own "Beginning
Farmer Real Estate Program" to first-
time land buyers with net worth less
than $150,000.)

Ohio Agricultural Finance Commission
65 S. Front Street, Suite 606
Columbus, OH 43215
614/466-2737

Oklahoma Department of Agriculture
Oklahoma Beginning Farmer Loan
    Program
2800 N. Lincoln
Oklahoma City, OK 73105
405/521-3864

South Dakota Department of
    Agriculture
attention Mr. Kevin Richter
523 E. Capitol Ave.
Pierre, SD 57501-3182
605/773-5436

Texas Agricultural Finance Authority
Texas Department of Agriculture
P.O. Box 12847
Austin, TX 78711
512/463-7686

Wisconsin Housing and Economic
    Development Authority
Beginning Farmer Bond Program
One South Pinckney St., Suite 500
Madison, WI 53701
608/266-7884

Some states also have linked deposit programs which deposit state funds in banks that make low-interest operating loans to farmers. A more detailed description of both the aggie bond programs and other state beginning farmer programs, "State By State Agricultural Loan Programs," is available from Communicating For Agriculture, P.O. Box 677, Fergus Falls, MN 56538. Phone: 218/739-3241.